I0123980

On Terrorism and the State

(Chapter X of *Remedy for Everything*)

By Gianfranco Sanguinetti

Translated from the French
by Bill Brown

Colossal Books
Brooklyn, NY

Published February 2014

Colossal Books
POB 140041
Brooklyn, NY 11214

Cover design: George Matthaei.

ISBN: 978-0-615-96302-0
Printed and bound in the USA.

Contents

Translator's Introduction

In early 1979, Gianfranco Sanguinetti was hard at work on *Rimedio A Tutto: Discorsi sulle Prossime Opportunita' di Rovinare Il Capitalismo in Italia* ("Remedy for Everything: Discourses on the Next Chances to Ruin Capitalism in Italy"),[1] which was intended to be a follow-up to his *Rapporto verdico sulle ultima opportunita di salvare il capitalismo in Italia* ("Truthful Report on the Last Chances to Save Capitalism in Italy"). Published in August 1975 under the pseudonym of Censor, the *Rapporto verdico* had been a tremendous success. Not only had it received very positive reviews in the Italian press and had sold a lot of copies, but it had also caused a major scandal. No one had suspected that Censor (allegedly a conservative member of Italy's ruling class) did not exist and that, Sanguinetti, an anti-capitalist revolutionary and a former member of the Situationist International, had written the book, which were facts that he revealed five months after it had been published.[2]

Sanguinetti had certainly been stung by the rebukes made of him in mid-1978 by his friend and collaborator, the ex-situationist Guy Debord, who had unsuccessfully encouraged him to go public with the truth about Aldo Moro while the Italian Prime Minister was still alive (allegedly kidnapped and murdered by the Red Brigades, Moro was in fact abducted and killed by Italy's intelligence agencies).[3] Perhaps Sanguinetti was also motivated by the fact that, in February 1979, Debord had published his *Preface to the Fourth Italian Edition of "The Society of the Spectacle,"* which in part discussed the Moro

---

[1] The title is a détournement of François-Joseph Lange de La Maltière's *Remède à tout, ou constitution invulnerable de la félicité publique* ("Remedy for Everything, or the Invulnerable Constitution of Public Happiness"), first published in 1793.

[2] See my translation of Gianfranco Sanguinetti, *Truthful Report on the Last Chances to Save Capitalism in Italy* (Colossal Books, 2014).

[3] See Debord's letters to Sanguinetti dated 21 April 1978 and 29 August 1978, published in Editions Champ Libre *Correspondance, Vol. II* (Paris, 1981), pp. 97-100 and p. 118, and in *Guy Debord Correspondance, Vol. 5, Janvier 1973 – Décember 1978* (Librarier Arthème Fayard, 2005), pp, 455-459 and p. 473. Sanguinetti's responses of 1 June 1978 and 15 August 1978 are included in Editions Champ Libre *Correspondance, Vol. II*, pp. 100-117. Moro was found dead on 9 May 1978, almost two months after he had been abducted.

affair. In any event, Sanguinetti decided to publish the tenth chapter of *Rimedio A Tutto* as a book in and of itself. Originally titled *Terrorismo di stato e stato di terrorismo* ("State Terrorism and the State of Terrorism"), this text was published in April 1979 under the title *Del terrorismo e dello stato: La teoria e la practica del terrorismo per la prima volta divulgate* ("On Terrorism and the State: The Theory and Practice of Terrorism Divulged for the First Time"). The first part of this new title was intended as an echo of *Del principe e delle lettere* ("On the Prince and Letters"), a revolutionary pamphlet written by Vittorio Alfieri in 1795.[4] The second part seems to be a dig at Debord, whose *Preface* had claimed to be the first text to speak truthfully about Italian terrorism: "Of these sad facts many Italians have been aware, and many more straight away took them into account. But they have never been published anywhere, because the latter have been deprived of the means of doing it and the former of the wish to do so."[5] This claim ignored the existence of Censor's *Rapporto verdico*, which had been published more than three years previously.[6]

Perhaps because it had been rushed into print, *Del terrorismo e dello stato* presented itself in a manner that was slightly confusing. The table of contents for *Rimedio A Tutto*, as well as Sanguinetti's various introductions to it (a "Notice from the Author," a "Dedication to the Bad Workers of Italy and All the Other Countries," and a "Preface"), accompanied it. But the rest of the book was never published and, as Sanguinetti relates in his preface to the French edition of his book, *Del terrorismo* "was not reprinted in Italy because of several difficulties created for me by a stupid and crude judicial-police persecution."

Del terrorismo e dello stato was quickly translated into French as *Du Terrorisme et de l'Etat: La théorie et la pratique du terrorisme divulgées pour la première fois* by two sets of translators: Jean-François Martos, a young man

---

[4] In 1989, Sanguinetti and Editions Allia published a French translation of this work under the title *Du Prince et des Lettres*. Email to me dated 2 October 2012.

[5] Guy Debord, *Commentaires sur la société du spectacle, 1988,* suivi de *Préface à la quatrième édition italienne de La Société du Spectacle, 1979* (Gallimard, 1992), p. 142.

[6] But even Censor/Sanguinetti had not been the first, a distinction that can only be claimed by the Italian section of the Situationist International, which published *Il Reichstag Brucia?* ("Is the Reichstag Burning?") on 19 December 1969.

going to school in Paris, and Jean-François Labrugère and Philippe Rouyau, two young men living in Grenoble and working as a team. In the first half of 1980, these men published their respective translations, both of which included a new preface that Sanguinetti had written in January 1980. But unlike Martos, who produced a second, corrected edition of his translation, Labrugère and Rouyau only produced a single edition of theirs. On 13 August 1980, they wrote to Gérard Lebovici, the editor-in-chief of Editions Champ Libre, which had published a French translation of Sanguinetti's *Rapporto verdico* in January 1976, and asked Lebovici if he would assist them in publishing a new edition of their translation of *Del Terrorismo*. Lebovici refused, in part because he didn't think very much of the quality of Sanguinetti's second book, and in part because he was offended by its subtitle, which ignored the fact that Champ Libre had published Debord's *Preface to the Fourth Italian Edition of "The Society of the Spectacle"* two months before Sanguinetti had come out with the Italian edition of his book.[7]

At the time that Martos published his translation of *Del terrorismo*, he wasn't one of Guy Debord's friends. But when the two men met in March 1981, Debord immediately pressed him to join his disinformation campaign against both Sanguinetti and his book. In a letter to Sanguinetti dated 4 April 1981, Martos says that Guy "isn't angry with you, but he has simply 'broken off relations.'[8] He thinks that this attorney, of whom you have spoken a bit to me,

---

[7] To read the letters exchanged between Lebovici and Labrugère & Rouyau, see Editions Champ Libre *Correspondance, Vol. II*, pp. 69-72.

[8] It appears that this decision to break off relations was reached in October 1978. Cf. Debord's letter to Paolo Salvadori dated 12 November 1978, published in *Guy Debord Correspondance, Vol 5*, pp. 482-485: "Thus I have telegraphed [Gianfranco], without explanation, that our meeting in Geneva has been canceled. As you know, I have shown him extraordinary patience on the personal level because he merits it for several reasons. And though I have interrupted all relations on that level for nearly three years, I would still like to think that there still remains a chance for him to manifest his talents in an autonomous manner in the general activity of 'our party.' The question can no longer be posed." The reason for this decision concerned Sanguinetti's manuscript, not his behavior or the people with whom he was associating: "Although there are several good pages and a generally acceptable intention, and certainly courage (if it is to be published soon in Italy), it is necessary to say that this book, when considered as a whole, constitutes an irreparable and monstrous disaster. Everything is lacking: in the strategy of the discourse, in

and who is surnamed 'the doge' – Mignoli? – is an officer in the secret services and that you should be suspicious of him."

Martos would write to Sanguinetti again, on 3 June 1981.[9] "I have recently received two documents that you already know: one is the correspondence between you and 'Cavalcanti,'[10] which Guy has made available to me and Michel [Prigent]. The other is Els van Daele's 'Postface to the Dutch translation' of *Terrorismo*," Martos wrote. "Given the critiques of you that are developed in these texts, *tacere non possum*, it is thus necessary that I give you my opinion of them, holding myself to the strict truth [...] As all of this is now discussed by several comrades, and so as to make precise to them what I think, I have also communicated this letter to them. And, awaiting your response, or better still hoping to see you if you come to Paris, I send you and Katarina my best wishes." Sanguinetti didn't respond to this letter, a fact that Debord interpreted as "a terrible verification: even more than I would have thought." According to him, the "quite polite tone of the questions that you posed to Gianfranco had the merit of allowing him complete freedom to respond and offered no excuse for a cop-out."[11]

---

the 'literary' construction of the text as a whole, in its very style, which is at once maladroit and pretentious in the extreme, in the figure that the author puts forth everywhere and that succeeds in being vividly antipathetic and, at the same time, completely ridiculous. To summarize the fundamental error of *the author*, one can say that he has, so as to surpass 'Censor,' stupidly reprised this glorious personality, with all of his idiosyncratic expressions, but in a *debased* manner because he has passed over to the side of the proletarians, with the result that the discourse takes on an aspect that evokes the beards of the old, autodidactic anarchists of the end of the 19th century. And to summarize the error of *the man*, it is necessary to say that the most lamentable sides of his personality, which once a month or so express themselves by inept comportment in a restaurant, are spread about without limits in the language of historical action."

[9] This letter was published in Jean-François Martos, *Correspondance avec Guy Debord* (Le fin mot de l'histoire, 1998), a book that was removed from circulation the following year after a successful claim of copyright infringement was lodged against it by Librairie Arthème Fayard and Alice Becker-Ho aka "Alice Debord."

[10] A reference to the letters Debord sent Sanguinetti on 21 April 1978 and 29 August 1978. Cavalcanti was the pseudonym that Debord had used in this correspondence.

Debord's behind-the-scenes campaign against Sanguinetti's *On Terrorism* continued into 1982. Daele's "Postface," which was either based upon materials that Debord had furnished or had been written by Debord himself, was followed by Lucy Forsyth's "Foreword to the English Edition,"[12] which was a simple reiteration of the contents of Daele's "Postface." In the words of Sanguinetti's letter to Mustapha Khayati, which appears at the end of this volume, these translations of Sanguinetti's *Del Terrorismo* "are the most striking examples of schizophrenia in the history of publishing since *Anti-Machiavel* by Frederic II and Voltaire." Both of them "publish my text and, at the same time, launch an attack against my person [...] This gives the impression that the book was only published so that their suspicions about and censures of its author could be spread." To make matters worse, Forsyth's translation is overly literal and full of typographical and grammatical mistakes. Until now, it has been the only translation of *On Terrorism and the State* available in English.

* * *

Though it was one of the very first texts to be published on the subject of terrorism in Italy during the 1970s, *On Terrorism and the State* is completely absent from "mainstream" discussions of the subject. The list of books in which it is not mentioned is truly extensive: Kenneth R. Langford's *An Analysis of Left and Right Wing Terrorism in Italy* (Defense Intelligence College, 1985); Leonard Weinberg and William Lee Eubank's *The Rise and Fall of Italian Terrorism* (Westview Press, 1987); Richard Drake's *The Revolutionary Mystique and Terrorism in Contemporary Italy* (Indiana University Press, 1989); Robert C. Meade's *Red Brigades: The Story of Italian Terrorism* (Macmillan, 1990); Raimondo Catanzaro's *The Red Brigades and Left-wing Terrorism in Italy* (Pinter, 1991); Marco Rimanelli's *Waning Terror: Red Brigades and Neo-Nazi Terrorism in Italy* (World Jurist Association, 1991); Jeffrey McKenzie Bale's *The "Black" Terrorist International: Neo-fascist Paramilitary Networks and the "Strategy of Tension" in Italy, 1968-1974* (University of California, Berkeley, 1994); Paul Ginsborg's *A History of*

---

[11] Letter from Debord to Martos dated 29 August 1981 and published in *Guy Debord Correspondance, Vol 6: Janvier 1973-Décembre 1978* (Librairie Arthème Fayard, 2005), pp. 178-180.

[12] Gianfranco Sanguinetti, *On Terrorism and the State: The theory and practice of terrorism divulged for the first time* (London: B.M. Chronos, 1982), pp. 6-13.

Contemporary Italy: Society and Politics, 1943-1988 (Palgrave Macmillan, 2003); Daniele Ganser's *NATO's Secret Armies: Operation GLADIO and Terrorism in Western Europe* (Routledge, 2004); Silje Dalsbotten Aass's *State Responses to Terrorism in Italy: The Period 1969-1984* (S.D. Aass, 2005); Graeme Allen Stout's *Arrested Images: Discourses of Terrorism in Italy and Germany* (University of Minnesota Press, 2006); Anna Cento Bull's *Italian Neo-Fascism: The Strategy of Tension and the Politics of Non-Reconciliation* (Berghahn Books, 2007); Pier Paolo Antonello's *Imagining Terrorism: The Rhetoric and Representation of Political Violence in Italy 1969-2009* (MHRA, 2009); and Richard Cottrell's Gladio, *NATO's Dagger at the Heart of Europe: The Pentagon-Nazi-Mafia Terror Axis* (Progressive Press, 2012), among many others.

It is possible that none of these books mention *On Terrorism and the State* because its author is virtually unknown outside of certain, very limited circles and because, over the years, copies of his book have been almost impossible to find. In the words of one of the very few authors who does refer to it – Philip Willan, the author of *Puppetmasters: The Political Use of Terrorism in Italy* (Constable, 1991) – Sanguinetti's book, which is described as "maverick," is "rare" and "privately published." That is to say, virtually no commercial distributor carries copies of it; it is only available through anarchist or informal distribution networks. And yet, according to WorldCat.org, which describes itself as "the world's largest library catalog," three libraries have copies of the Italian original;[13] thirty-four have copies of Martos' translation; three have copies of the translation by Labrugère and Rouyau; forty-five have copies of the English translation; six have copies of the Dutch translation; eleven have copies of a German translation;[14] one has a

---

[13] None of these copies are housed in Italy. The only library in Italy that has a copy of the book is the Johns Hopkins SAIS Bologna Center, which has a copy of the English translation.

[14] This translation was published in 1981 by Edition Nautilus, a publishing house that, according to Martos' letter of 4 April 1981, was to be distrusted because "their translations are bad, without mention of origin, their catalogue contains anything and everything and, bizarrely, though they constitute a certain pole of attraction in Germany, they never seem to have enemies among the Teutonic police. . . . And, without affirming anything with certainty, one could relate their fetishism of organization to the quasi-cop [*quasi-flicarole*] letter that they sent to Michel Prigent."

copy of a translation into Greek; and one has a copy of a translation into Spanish.

What about the Internet? Ever since 1999, my website (www.notbored.org) has hosted Lucy Forsyth's translations of the prefaces that Sanguinetti wrote to the Italian and French editions of his book, and, ever since 2004, it has hosted her translation of *On Terrorism* itself.[15] But with a handful of exceptions (see below), the Internet has paid virtually no attention to Sanguinetti's book. For example, no mention of *On Terrorism and the State* is made in the Wikipedia entries for "Operation Gladio," "Gladio in Italy," "the strategy of tension," "the years of lead," "false flag terrorism" and "state terrorism." Nor is Sanguinetti's book mentioned in any of the many articles devoted to terrorism, the strategy of tension, and Italy in the 1970s that are archived by libcom.org, a website devoted to and administered by adherents of libertarian communism.

There is nothing new about this silence. In January 1980, in his preface to the French edition of *On Terrorism and the State*, Sanguinetti himself notes the existence of "the quasi-complete silence that has surrounded a book that deals with a subject that is spoken about every day, but always in the same mendacious way, on the front pages of all the Italian newspapers as well as on the State-sponsored radio and television stations" and notes that the existence of his book has been "kept secret by the very people who are believed to have the obligation to speak about terrorism." The reason for this silence is, I believe, easy to imagine. Sanguinetti didn't simply assert what many people had refused to believe at the time, namely, that the Italian State had bombed, wounded and even killed some of its constituents, and had cynically blamed others for these crimes. He also denounced those who, through either stupidity or self-interest, adamantly refused to believe that such a thing could *ever* happen. And these people, and those for whom they spoke, never forgave him, even though – or precisely because – history has proved that Sanguinetti was right.[16] Such is the price for proving that the experts have lied: they lie about you; they deny that you even exist.

Among the exceptions is a man named Webster Tarpley, who is the author of *Synthetic Terror: Made in USA* (Progressive Press, 2005). It is clear from his footnotes and bibliography that he encountered Sanguinetti's book through my website. Not only does he mention *On Terrorism and the State*, but

---

[15] This was before I translated all these texts from scratch.

[16] Cf. Daniele Ganser, *NATO's Secret Armies: Operation GLADIO and Terrorism in Western Europe* (Routledge, 2005).

he also quotes from it extensively (sometimes with proper attribution, sometimes without it). To him, Sanguinetti's book offers support for the idea that Al Qaeda didn't perpetrate the attacks carried out in the United States on September 11, 2001 – the CIA did. Sanguinetti agrees with this thesis, but, unlike him, Tarpley is not a libertarian communist. In fact, he is an anti-Communist zealot and a bit of a lunatic. For example, he thinks that the CIA created and financed the Situationist International.[17] Does he know that Sanguinetti belonged to the SI between 1969 and 1972? If Tarpley were told about it, would he think that Sanguinetti's membership in the SI somehow undercuts the validity or usefulness of his critique of the CIA? I don't know. It doesn't appear that anyone has ever asked Tarpley these questions. But I have read enough of his writings to make an educated guess about how he would respond if he were told that an ex-situationist had written a book that denounced the CIA's machinations. He would call that book a "limited hang-out operation"[18] and then claim that he was never fooled by it, not even for a second.

The other exceptions are those people who also believe that the CIA was behind the attacks that took place on September 11, 2001 but who, like

---

[17] Cf. interview with Webster Tarpley, Press TV, 13 October 2011: "[The] Situationist International was cooked up by NATO and the CIA back in the 1950s and 60s to overthrow General de Gaulle of France who was the target at that time."

[18] This is an operation "in which carefully selected and falsified documents and other materials are deliberately revealed by an insider who pretends to be a fugitive rebelling against the excesses of some oppressive or dangerous government agency. But the revelations turn out to have been prepared with a view to shaping the public consciousness in a way which is advantageous to the intelligence agency involved. At the same time, gullible young people can be duped into supporting a personality cult of the leaker, more commonly referred to as a 'whistleblower.' A further variation on the theme can be the attempt of the sponsoring intelligence agency to introduce their chosen conduit, now posing as a defector, into the intelligence apparatus of a targeted foreign government. In this case, the leaker or whistleblower attains the status of a triple agent." Webster Tarpley, "How to identify a limited CIA hangout op?" Press TV, 18 June 2013. The reactionary nature of this obscurantist analysis can be seen in the fact that, for Tarpley, Daniel Ellsberg, Julian Assange and Edward Snowden are not genuine whistle-blowers, but "triple agents."

Sanguinetti, are libertarian communists. A pair of them (Jeff Strahl and Tod Fletcher) have uploaded my translation of *On Terrorism and the State* to a blog called the "Daily Battle"[19] and have added footnotes that show the many parallels that exist between the terrorist attacks carried out in Italy in the 1970s and the attacks perpetrated on September 11, 2001 in the United States. Those parallels include the following:

(1) Both sets of attacks were preceded by predictions that public opinion about the pressing issues of the day would not change unless some kind of major catastrophe took place.

(2) Both sets of attacks were preceded by events that embarrassed the State (the inability of the unions and police forces to contain working class rebellion during 1969 in Italy and the success that anti-globalization protests had in Seattle in 1999 and Genoa in 2001).

(3) Both sets of attacks were never claimed by any individual or group, but were quickly blamed on extremists.

(4) Despite their limited means, these extremists were able to perpetrate spectacularly successful attacks against much stronger adversaries.

(5) Both the Red Brigades and Al Qaeda were manipulated, if not actually created, by the intelligence agencies of the countries that were attacked by them.

(6) Both sets of attacks were used as justifications for the quick passage of legislation that had been drafted long before these attacks and were used to criminalize completely legitimate forms of protest.

---

[19] http://www.dailybattle.pair.com/2013/sanguinetti_state_terror.shtml. This excellent blog also hosts like-minded essays by Max Kolskegg ("9/11 in Context: Plans and Counterplans" and "9/11: A Desperate Provocation by US Capitalism") and an interview with Tod Fletcher ("9/11 in Context: The Strategy of Tension Gone Global").

(7) Left-wing intellectuals were quick to believe and repeat the State's statements about the identities of those who had perpetrated the attacks and to denounce those who didn't believe these statements as "conspiracy theorists."

Though I am sympathetic with these efforts or, rather, though I agree that these parallels are significant, I don't believe that this analysis gets to the heart of the matter.

First and foremost, while Italian capitalism was experiencing a real crisis in the late 1960s and early 1970s (its working class was not only rebelling, but was also rebelling in a truly radical and quite effective fashion), American capitalism in 2001 was not. Questions about the legitimacy of the election of a particular president are not questions about the legitimacy of the system as a whole. Furthermore, protests against meetings held by the World Trade Organization or the International Monetary Fund, even when they are massive, are quite anodyne in comparison to the sabotage of industrial production and participation in wildcat strikes. Second, while Italian capitalism officially proclaimed that it was menaced by and was fighting back against anarchist and Communist subversion (that is to say, something that threatened the country's class structure), American capitalism officially proclaimed that its enemy was Islamic fundamentalism (something that threatened its religious identity and its "democratic freedoms"). Third and last, Italian capitalism was defending itself with a weapon – Operation Gladio – that had been forged more than twenty years previously. The attacks carried out on September 11, 2001 took many, many years of planning; they certainly weren't set in motion just two years before they took place.

But this doesn't mean that people like Strahl and Fletcher aren't on the right track or that *On Terrorism and the State* isn't relevant to a critical analysis of September 11 and other instances of spectacular or artificial terrorism. In fact, it seems that there are more than mere "similarities" or "parallels" between Italian terrorism in the 1970s and the "global war against terrorism" that was launched in response to the attacks of September 11. They are, it seems to me, part of one and the same operation. In the words of one commentator, the attacks of September 11 and the subsequent global war against terrorism were part of "Gladio B": that is to say, the continuation and expansion of what the CIA and NATO were doing in Italy and the rest of Europe in the 1970s.[20] The central players are the same: one need only call

---

[20] Dr. Nafeez Ahmed, "Why Was a *Sunday Times* Report on U.S. Government

attention to the continued presence of Henry Kissinger to see this.[21] The justification is the same: the State needs to guarantee "continuity of government"; it needs to have the same people in command, even if the thing that threatens that continuity has apparently changed (it used to be a Communist takeover, now it is a terrorist attack). And the ultimate goal is the same: control of the world's supplies of oil and natural gas. The only difference is that while "Gladio A" used neo-Nazis to fight against the Communists in places like Italy and Belgium, "Gladio B" uses *mujahideen* to fight against the Communists in Afghanistan and the Balkans. In sum, the Cold War never ended; it simply entered a new phase.

If *On Terrorism and the State* is relevant today, almost 35 years after it was first published, this is because of its author's commitment to the importance of historical knowledge and to seeing the continuity "behind" or "between" apparently unrelated or unprecedented events. The perpetrators of the attacks that took place on September 11 have been successful in their attempts to capitalize on those attacks because they have managed to convince people that, on that day, "everything changed." It is only a detailed knowledge of history that allows us to see that, no, "everything" didn't "change" on that day. In point of fact, "everything" remained very much the same: the rich and powerful remained in control, and they continued to want to make sure that they never lose their wealth, their power or their ability to control others. In fact, it is precisely change that they fear; they are especially fearful that, one day, "everything" might actually change. Of course, change is inevitable; it is impossible to forestall change forever. This is precisely why the rich and powerful are so dangerous. They grow more desperate every day.

* * *

A few notes about the text and the book's design. Since I cannot read Italian, I have used Jean-François Martos' *Du Terrorisme et de l'Etat: La théorie et la pratique du terrorisme divulgées pour la première fois* as the basis for this translation into English. I have dropped the always controversial and now increasingly irrelevant subtitle. The original Italian edition included words and phrases from a number of other languages (mostly Latin, French and

---

Ties to Al Qaeda Chief Spiked?" *Ceasefire*, 20 May 2013.

[21] Other players were in power during both the early 1970s and 2001 include Dick Cheney and Donald Rumsfield. Cf. Kevin Ryan, *Nineteen 9/11 Suspects* (Create Space Independent Publishing Platform, 2013).

English). Martos was careful to preserve this multi-lingual richness as he translated the work as a whole from Italian into French, and I, translating from French into English, have tried to be careful, too. When Sanguinetti quoted from an Italian translation of something in English, I sought out and used the original wording. When he quoted from something in Latin, I consulted and relied upon the already-established rendering of it into English. All of the footnotes are by me, except where noted. Both Els van Daele's 'Postface to the Dutch translation' of *Terrorismo*" and Sanguinetti's letter to Mustapha Khayati have never appeared in print or in an English translation before. Finally, this edition of *On Terrorism* is the first one to include an index of the important names, events and places mentioned in the text.

Bill Brown
Brooklyn, NY
1 January 2014

On Terrorism and the State

Notice From The Author

Today, those who fear ideas have little fear of books. Each week the market offers an infinite number of books and no ideas; people seek their ideas outside of the market and the bookstores. And in Italy as in Iran,[1] people find what they seek in the streets.

Everything leads one to believe that, if thinking in the form of writing still is not prohibited in our country, this depends less upon the liberalism of the legislators than the fact that one does not run the risk of reading anything new, with the result that those who want to read books that are worth the effort *must write them themselves,* given that this sector of social production is (as much as the others) subject to the currently prevailing conditions of falsification and pollution. The very editors who today publish *anything* do indeed publish anything and, seeing what they publish, we can be certain that we can find the most interesting things among what they do not publish. Here I can provide easy proof, without which one might think that it is due to a lack of interesting writings that Italian editors do not publish anything interesting.

For two years after the success of the scandalous pamphlet that I wrote under the pseudonym Censor,[2] several bourgeois publishers informed me that they were completely disposed to close their eyes to the subversive content of what I wrote so that they did not have to deprive themselves of the business that, in their opinion, my publications could bring them. When I committed myself to writing another book, one entitled *Remedy for Everything,* Arnoldo Mondadori hurried to submit to me a publishing contract according to which, in addition to the book in question, he would have a monopoly on all of my publications for *the next ten years,* which was obviously a pretense that was beyond his means; consequently, I refused. As a result, he limited himself to paying for the book in advance, which he bought, as one says, *with his eyes closed to the risks and dangers.*[3]

But when the zealous directors of this famous publishing house read the completed manuscript, they were literally *terrorized,* as if no one could still *show in writing what could be said* of this [Italian] State and its entire spectacle. According to the specialists in *marketing,*[4] subversive ideas can

---

[1] On 16 January 1979, after two years of intense demonstrations and protests, the Shah (Mohammad Reza Pahlavi) fled Iran, which he had ruled since 1941.

[2] *Truthful Report on the Last Chances to Save Capitalism in Italy* (1975).

[3] *a scatola chiusa.*

[4] English in original.

1

certainly sell well and, in any case, better than an absence of ideas, the sale of which is the specialty of those gentlemen. But at a time when the workers no longer want to be workers, we should not be too surprised to see that publishers fear being publishers. In this case, we can say that these audacious *managers*[5] not only bought my book with their eyes closed to the risks and dangers, but also *thoughtlessly and casually*,[6] as one says in Florence, because they must have and did imagine that I would write either a eulogy for this world or a futile lament for it. They hoped to do good business with subversion and, instead of that, they lost their ill-advised investment of "venture capital" by paying [me] so as to not to buy [the book from me]! These incapable but, in the final analysis, entertaining managers of this publishing house resemble the managers of our entire bankrupt economy. No one should be surprised if, very rapidly and not only thanks to the merits of these *managers*,[7] that economy ends up in the most complete ruin.

Waiting to *make a little Iranian revolution,* and certainly a better one, here among us, I only publish (for the moment) the dedication and the preface from *Remedy for Everything* conjointly with the chapter in it that relates to the terrorism against the proletariat that our State has practiced with impunity for the last decade. As for the rest of the book, it can wait. Nevertheless, the truth about terrorism, which one can immediately read here *and only here,*[8] has no publisher, but as the reader will see, it has no need of one. This truth violently refuses the clandestinity to which one would like to reduce it, and it is capable of inaugurating an *Italian samizdat* to get itself diffused.

Today, the innumerable enemies of that truth – who come from the Center, Right and Left – *must declare themselves* by exposing themselves on uncovered ground to combat it, because all of their lies no longer manage to hide it. And no matter what one can say about that truth today, in five or 10 years, or even before then, when everything has become clear to everyone it will be what I have written about terrorism that one will remember, and not the streams of ink that all the professional liars and the imbeciles currently

---

[5] English in original.
[6] *a bischero sciolto.*
[7] English in original.
[8] This line greatly displeased both Gérard Lebovici and Guy Debord, who complained that Sanguinetti had ignored the existence of Debord's *Preface to the Fourth Italian Editions of "The Society of the Spectacle"*, which Editions Champ Libre had published in February 1979, two months before the publication of *Del Terrorismo E Dello Stato.*

release on the subject.

To those who fear the truth, I want to offer several truths that will make them fearful; and to those who have no fear of it, I want to offer a reason that proves that the terrorism of the truth is the only one that benefits the proletariat.

Milan, March 1979

"I know that you will not make yourself an accomplice to an operation that, more than all the others, would destroy C[hristian] D[emocracy] (...) The first thing to say is that it is a matter of something that repeats itself (...) One speaks of it less these days, but enough for you to know how things take place and you, *who know everything,* are certainly informed of it (...) But (...) to spread calm in the entourage (...) you can *immediately* appeal to Pennacchini, who knows everything (in detail) better than I (...) Then there is Miceli and (...) Colonel Giovannoni, who Cossiga esteems (...) After some time has passed, public opinion will understand (...) The important thing is to convince Andreotti that, if he plays the trump card, a bloc of intransigent opponents will probably be constituted."

(Aldo Moro, letter to Flaminio Piccoli, only made public on 13 September 1978)

"I know that the demand for an intransigent truth has spread. But I also know that many things (...) require discretion, silence (...) And this in the interest of the objectives that we want to attain. This is exactly why, since the day of my arrival at this Ministry, I have not ceased to recall to each person the duty of discretion, I would even say the wisdom of silence."

(Virginio Rognoni, Minister of the Interior, 24 August 1978)

"When fate decrees that the people will have faith in no one, as sometimes occurs, after they have been deceived in the past either by events or by men, this inevitably leads to ruin."

(Machiavelli, Book I, Chapter 53, *Discourse on Livy*)

Remedy for Everything:
Discourses on the Next Chances for Ruining Capitalism in Italy

Table of Contents

Gianfranco Sanguinetti

Dedication to the Bad Workers of Italy
and All the Other Countries

No doubt it is not yet time to do good. The particular good that we make is [merely] a palliative. We must await a very great, general evil for general opinion to prove its need for proper measures that do good. *What produces the general good is always terrible or appears bizarre when we begin too soon.*

(Saint-Just, *Posthumous Writings*)[9]

It is to you, the bad workers, that I address this pamphlet, which, if it does not answer the obligations that I have to you, is nevertheless the greatest gift that I can send you in these times, because I have sought here to express in words the same total, resounding and salutary insubordination that you express even better and always with more radicalism through your actions and your struggles against work. And since neither you nor the others can wait another hour for me without contenting yourself less as a result, you must not complain that I have not given you more than this. Perhaps you should criticize me for not having known how to describe all the misery (which is very great) against which you revolt today or for not having known how to reveal all of the richness of your revolt, which is not slight, but in such case I do not know who is less obligated to the other: me to you, because you have encouraged me to write what I could not have written without you, or you to me, because by writing this pamphlet I will not have satisfied you.

Thus, take hold of this *Remedy for Everything* as one takes everything that comes from a friend, always considering the intention of the giver more than the quality of what one receives from him.[10] And my intention, just like yours, is to harm this world, which harms you, to unmask those who are paid to deceive you, and to take away the good reputations of those who still enjoy them. Nevertheless, if here I directly attack the men who are known today, but who will quickly be buried by oblivion or by the consequences of their own abuses, it matters less to me to displease them than, through them, to strike

---

[9] Louis Antoine Léon de Saint-Just (1767-1794). Fragment #3, *Fragments on Republican Institutions*.

[10] The first four sentences of this text are closely modeled on the "Greetings" that begin Book I of Machiavelli's *Discourses of Livy*.

against *all the institutions* of this society, institutions that they represent so well but defend so poorly, in the hope that they will be defended in their turn. My only desire is that this pamphlet will be capable of inciting those who still work without protesting – the good workers – to be *less good,* and those who, like you, already revolt – the bad workers – to become *even worse.*

To write such things against this world is easier than reading them, and reading them is easier than doing them. As for me, I would prefer to read what I write and to see and do what I read. Despite this, I would consider myself *hardly practical* if today I did not put my pen to certain ends *a little better* than so many others who say they use weapons and in a manner that I believe is less efficacious, because it is the pen that puts weapons to work, and not the reverse, as is desired by the owners of this society and the naïve fanatics of armed struggle, who, on this point, are more in agreement than they would like to believe.

If you, the bad workers, judge that this *Discourse* is not too inferior to the ambitious intention that animates you and animates me, I will not fail to do even worse the next time, pushed as I am by my natural desire to commit (without any respect) everything that could *bring the attack* to the masters of our world, our times and our lives. If, moreover, you find in these pages only a single supplementary reason to unleash new and more violent attacks against all those who oppress and exploit you – the bureaucrats and the bourgeois – and to violently demystify the hoaxers who still pretend to speak in your name and place, *Remedy for Everything* will have satisfied my desires and there is nothing I desire more than that.

Gianfranco Sanguinetti

Preface to the Italian Edition of *Remedy for Everything*

Victory will go to those who know how to create disorder without loving it.

(Guy Debord, *Internationale Situationniste* #1, 1958)

Intelligence is perhaps the most widely shared thing in our country: each person thinks that he or she is so well provided with it that the very ones who are ordinarily the least likely to find contentment in any other thing, like our leaders, are not in the habit of desiring to have more of it than they already do. And since it is not likely that everyone is deceived about this subject, we must wonder how and by what necessity or mysterious interest this intelligence, allegedly possessed by such a large number of people, appears so infrequently in our country and not at all, not even on exceptional occasions, among those who, either in power or seeking to gain access to it, continually tell us that, if they are incapable, it is our fault and that, if Italy comes to ruin, the fault won't be theirs.

The fact is that this country, which proclaims itself to be free and democratic, is in fact led by several hundred heroic imbeciles who fear the consequences of the intelligence of all the others much more than they fear the consequences of their own stupidity, who curb that intelligence by all available means to better give free rein to their stupidity, and who can get away with it because their stupidity does not risk being publicly sanctioned by our sporadic electoral circuses, while they nevertheless make daily use of it according to their whims. In such a social and political organization, which these gentlemen have so opportunely fashioned in their own image, it seems completely normal to me that any voice that distinguishes itself from the dominant mediocrity and makes no compromises with it is naturally reduced to silence by a multitude of quasi-automatic mechanisms that perhaps remain the only relatively efficient things amidst the general lack of efficiency.

For my part, I have never presumed to be more perfect than anyone else. On the contrary, I have often desired to have the quick and lively intelligence and imagination of someone else. I have only had the chance to be engaged, from a young age, in a voyage down a road along which I have encountered a few people of the highest intelligence that this era has produced, despite itself, and I have no fear of admitting that this has already permitted me to *harm* this world, that is to say, its owners, not as much as I

8

would have desired, but certainly more than the modesty of my forces, on their own, would have allowed me to hope.

Naturally, I do not exaggerate those first results because I have not been contented with them, just as I know that no one would be so unjust as to attribute to a single person, or to several people, the fault or the merit of having thrown our class society into a war in which the multi-colored forces of preservation have found themselves on the defensive and in an always-more precarious way. Those who first contributed to this, apart from the fact that the historical circumstances were favorable, were innumerable young proletarians who – although they are not known by name – remain the principal protagonists.

Without fear of being contradicted, I can even affirm that the last 10 years of class struggle have already allowed us to harvest such results and have showed so clearly the incapacity and abjection of our enemies – both bourgeois and Stalinist – that we cannot fail to consider the recent progress of the subversion of the dominant order with extreme satisfaction. Indeed, we can expect so many further encouragements in the future that, if this subversion (among the occupations of men) is the one that is serious and has a solid future, I dare to believe that it is the very one that I chose in an era in which certain choices were less propitious than they are now.

Working *against this world* by obtaining tangible results, that is to say, by not being contented with the principal ideological compensation of being part of an impotent "opposition," is a long-term task that also involves some inconveniences. But to work *for this world* is not much easier and more than often becomes, both objectively and subjectively, quasi *impossible,* and here I am not only thinking of the new *selective* unemployment into which our bankrupt capitalism has thrown an entire generation of young proletarians, which is an action that testifies to an imprudence and a lack of foresight whose consequences have still not been fully measured. In reality, the question surpasses our frontiers as much as the crude errors of our politicians and economists do. All the allegedly "very serious problems of our times" derive from a single, very simple fact; for each and for all, *it is time to resolve all the problems* and to resolve them directly, by oneself and also collectively.

That this is in fact *possible* is demonstrated by the terror that this blunt perspective is capable of provoking among all the current bosses of alienation and their political and labor union servants. That this is now *necessary* and urgent, as well, has, on the contrary, no need of particular demonstration because our class society, which is already *essentially* uninhabitable, has now *visibly* become so. Those who cannot understand this must also give up the idea of understanding the rest.

The politicians, economists, psychologists, sociologists, semioticians, intellectuals, specialists in public opinion and all the other imbeciles who whore around with power ceaselessly evoke those "very serious problems" and yet keep to themselves what they really *designate.* Those who drool and quiver with delight each time that their bosses ask them to sniff out a new phenomenon in which the crisis manifests itself – that is, those who love definitions and labels – today find a thousand pretexts for never naming what their science can never resolve but which *they do not want to see resolved by other people.* In reality, their trade principally consists in *showing that they are necessary* to their employers, and this is precisely their dominant preoccupation in this period, when the proletariat thinks that neither these experts nor their bosses are necessary. If such a phenomenon seems curious, we can certainly not say that this is what determines the novelty of this era, because it is only a consequence and not even the most interesting one, and if there is something surprising in the phenomenon of general disarray, it is only the extravagant credit that such specialists continue to benefit from among those who continue to employ them in the hope of . . . we don't even know what. Here, as elsewhere, they confirm the old adage: the servant takes after his master.

Faced with such a tableau of the decomposition of the old world, the false consciousness that still reigns but *no longer governs* shamelessly accuses the young proletarian generation – which has re-launched the offensive against the society of the spectacle – of not being in a position to resolve the questions that are at the origin of its revolt and at the root of the crisis about which all the constituted powers are now debating. The contrary is what is true, because, in reality, the young proletarians are accused *of posing questions that power cannot resolve* from the moment that power itself is put into question.

And what do these famous "problems," neutralized or falsified by all the enslaved thinkers, really consist of? What are they precisely? Society divided into classes, work, property, the very conditions in which we are forced to survive and produce (as well as all that we are forced to produce and consume), the lies of bourgeois "democracy" and "liberty" (as well as the bureaucratic lies of "communism" and "equality") – in sum, the society of the spectacle in its entirety – *no longer functions* at the very moment that its reality *has been universally put into question* and attacked by a refusal that is not momentary and partial, but permanent and total.

All proletarians have been able to determine, at their own expense, that working for this world simply means exchanging one's life and time for a miserable salary that only guarantees survival and a perpetually precarious situation. And it is precisely salaried work that is being put into question and ultimately refused in a thousand different ways and on a thousand different occasions. More of a dialectician than his boss, the Italian worker is rediscovering a truth that old Hegel candidly expressed, but without really weighing the consequences or foreseeing the results: "Working means annihilating the world or cursing it."

Until now, the workers have limited themselves *to cursing this world*; today it is a question *of annihilating it.*

Ten years ago, "Never Work!" was written on the walls of Paris during the May revolution, and, in February 1977, this same command reappeared on the walls of Rome, greatly reinforced by the simple fact that, in the meantime, it had been translated into Polish by the workers in Stettin, Gdansk, Ursus and Radom in 1970 and then again in 1976, and also into Portuguese by the workers in Lisbon in 1974.

Surpassing the economy is the agenda everywhere, and proletarians, by refusing to work, show that they know perfectly well that work is principally *a pretext* that continually keeps them *under control* by forcing all of them to always be occupied with something other than their true interests. "On their banner, they must erase the *conservative* slogan 'A fair salary for a fair day of work' and write the *revolutionary* one 'Abolish salaried work!'" (Marx). Furthermore, even Lord Keynes, writing in his famous *Treatise on Money,* had to agree that "for anyone looking towards the future, the economic problem is not *the permanent problem of the human species*" – and in this he showed himself to be less obtuse than his current epigones and fervent out-of-season zealots. The fundamental fact is not so much that all the *material means* exist for the construction of free life in a classless society, as is the case today; it is, rather, that "the blind underemployment of class society's means can neither be interrupted nor go much further. Never has such a conjunction existed in the history of the world" (Debord).

I know several workers who are much more seriously occupied with [the reality of] political economy than unfortunate Franco Modigliani, and with more effectiveness than inept Giorgio Napolitano, but *from the opposite perspective*: that of the destruction of political economy. They put their theoretical discoveries into practice, and their critique of the economic system surpasses and invalidates the critique that unjustly famous Piero Sraffa believes that he has made. And, inversely, these workers are beginning to theorize the first practical results of their direct experiments *on the fragility of the economy.* They read Paul Lafarge's pamphlet *The Right to be Lazy,* in which – although it was written at the end of the 19th century and has been ignored by our ignorant economists – assuredly remains the most important and most modern work of pure critique of political economy to be appear after Marx. Well in advance and with great lucidity, Lafarge foresaw the reasons that capitalism would be led to [inaugurate] modern consumption, as well as the salient characteristics of what he calls "the era of falsification," which we are living in today. He also indicates the irremediable contradictions in such consumption and, finally, that which summarizes and resolves them: the refusal of work and the surpassing of the economy.

The workers have finally been forced to see that the colors in which the dominant spectacle arrays itself to camouflage its monstrous traits are the very same fatal colors produced by the *cancer factory* in Cirié – a factory that, as everyone knows, destroys the workers at the same time that it produces dyes. This factory can be justly cited as the admirable quintessence of all the others; the only difference between them is that the *destructive cycle* of its productive forces is slightly more rapid and radical than elsewhere. But all factories have a close relationship with the cancer factory.

* * *

As was said of Louis XVI, capitalism *must reign or disappear.* But to reign it must know how to constantly foresee and seek to avoid the breaking point of the unstable equilibrium that exists between everything that capitalism must impose and inflict on everyone – renunciation, sacrifice, constraint, boredom, pollution, et. al – and that which everyone can objectively support and is subjectively disposed to tolerate. Today, the very development of capitalism is such that, while the threshold of toleration *tends to fall* – as much for historical reasons as for simple biological ones – the quantity of all that this type of society must impose on us for its own particular necessities for survival *tends, on the contrary, to rise* without limits or wisdom, that is to say, due to its own movement, which is absolutely autonomous and independent from the real needs of men and women, and even from their most basic and irreducible requirements for survival. The spectacular-commodity society – that immense *immobile motor* – must constrain every person to sustain it and defend its anti-historical immobility. Nevertheless, the Herculean Columns of alienation, the limits that no one must ever cross, are no longer far away, or at the antipodes of the world or human knowledge, but are close to everyone, wherever they are. And every person must be capable of surpassing them if we do not want "to deny [the] experience, following the course of the sun, of that world that has no inhabitants" (Dante), that is to say, the experience *of the negative at work*, which is already the *practical* negation of all the limits that are arbitrarily imposed on the vast majority of humanity, on the proletariat, forced to live in a mindless state without ever giving any reality to its talents, its mutilated capacities or its unrecognized desires.

Descartes said: "My third maxim is to always seek (...) to change my desires rather than the order of the world." Today, since the times have changed – and, with them, men and women, and their aspirations and desires – we must abandon all uncertainty and scruples. And thus our first maxim reverses that of the philosopher: *Always seek to change the order of the world instead of changing your desires.* And the proletariat must now seek to conquer, not fail, because only a violent desire for victory can assure the victory of the most authentic desires, which are also the least confessed.

All of the developed industrial world now presents itself as a sinister and endless suburb, of which Cirié, Seveso and their surroundings are both *the anti-historical center* and *the image of its future,* that is, if this world remains any longer under the direction of those who proclaim themselves "responsible" for politics and the economy. And modern spectacular capitalism can already contemplate itself – as if it were looking into a magical mirror that reveals the near future – in the generally censored images of the monstrous children recently born in Seveso.

Our bourgeois philanthropists might regret that this is so, but soon they will regret even more *that things are no longer this way,* because the quantity of all that this society imposes and inflicts on us has already surpassed the threshold beyond which any barely maintained equilibrium has been violently broken and can only be violently reestablished, but *always more provisionally than before.*

In such conditions, where the development of class society in all its bourgeois and bureaucratic variants is opposed, not only to the interests of the vast majority but also to the most elementary and fundamental conditions for the simple survival of the species and individuals (as well as their will to live), it is not a question of the proletariat delaying or avoiding a social war that *has already begun,* nor is it a question of the proletariat exhausting itself in a multitude of small skirmishes that are ceaselessly renewed because they are ceaselessly condemned to failure and fought "for the defense" of some nonsense ("Salaries, Jobs, the Country," as the unionized and Stalinist scoundrels like to bark). On the contrary, it is a question of the workers counter-attacking by going on the offensive and winning the war in the entire theatre of operations, which is global, as is the current crisis of all the powers. Because what is in play today is nothing other than *the destiny of the world.* Nevertheless, it is not at all in the name of some so-called "historical mission" (more or less unavoidable and prophesized) that the proletariat is called to become *the class of historical consciousness,* but because it is only from the position of fundamental superiority that the proletariat can successfully attack and combat all the forces of *unconsciousness* that are represented "democratically" (and they are the only ones that are represented) in contemporary capitalism. Henceforth, these forces will principally manifest themselves through their failures, disasters and infamies.

Since its beginnings, capitalism has been combative, and for a long time it has fought against all the other retrograde forms of power and social organization that have been opposed to its expansion. It has imposed itself and come out victorious from the wars that it has fought because (and only to the extent that) its development and conquest have corresponded to historically determined necessities and possibilities, of which none of its ideologues have ever been truly aware, just as today none of its contemporary ideologues are aware of the fact that the historical task of capitalism has *ended.* Today, now that it has conquered the world, become exhausted by its very successes, and been managed in an insane way by the half-witted heirs of the conquerors of the past, capitalism must once again and above all confront precisely that which has permitted it to attain such power: the proletariat. The social peace that capitalism has enjoyed for so long – since the failure of the social revolution in Russia and all of Europe – has almost made it forget the existence of its old enemy and this at the same time that there is no doubt that, today, capitalism has completely lost its former combativeness. All of its efforts now aim at preventing a social war for which it is not prepared, which it is *already desperate to win,* but for which its preceding development – so exalted until recently – has created all of the presuppositions.

On the contrary, the proletariat always finds itself at the center of a daily and permanent conflict, which is sometimes open, most often hidden, but always violent, and has lasted for a century and a half. Today, the class that has been continually at war against the conditions of its own oppression must necessarily perish or gain the upper hand over all the other classes that – sometimes at war, sometimes at peace – are never as ready to attack, nor as prepared to defend themselves. On the other hand, it is in the very nature of this war that the property-owning classes can never annihilate their enemy, that is to say, *abolish* the proletariat, which would mean abolishing *the very conditions for their own supremacy.* They need the proletariat; the proletariat *has no need* of them. This is the heart of the matter.

As if all this wasn't sufficient, we must note that the logic of such a conflict also includes the fact that, while the property-owning classes are constrained to consider each of their victories as *provisional,* and each respite that the proletariat concedes to them as uncertain, the proletariat, for its part, is obligated by its very condition to never accept any peace *if it is not the peace of the victor.* And it is precisely this fact that makes the proletarians always increase their *immense pretentions* as they go along and despite their past defeats, which were also provisional. Thus, the workers of the entire world continually plunge into the most profound despair and, in a rhythm that always quickens, the same forces that are opposed to them just barely win their victories. It is precisely in this manner that the proletarians impose on themselves the higher necessity of winning, not only this or that particular battle, but *the entire war.*

* * *

Marx said that human beings only pose the problems that they can resolve, and I would add that, today, we have precisely arrived at the point where *it is no longer possible to resolve any of them without resolving all of them.* This is why this pamphlet is called *Remedy for Everything.*

Our strength exactly resides in the facts that we are faced with *all the problems* and that we also have the necessity, as well as the possibility, of *resolving them all.* By contrast, the weakness of our enemies – bureaucrats and bourgeois – lies in the fact that, in addition to being confronted by all the problems, they have the imperative necessity of *not resolving them all,* that is to say, they are in a situation in which they truly cannot resolve any of them. Thus, this is exactly the nature of their situation today: they do not have the strength to resolve any problem and yet they are not in a position to prevent others from resolving them, nor can they coexist with these problems for very long. Thus, we must not be surprised by the fear and confusion that will henceforth reign in their ranks.

Until 10 years ago, it seemed impossible to the greatest number of people that anything could be changed; today it seems impossible to everyone *that anything can continue as it did before.* And yet, 10 years ago, the resigned thinkers of the impotent Left pompously decreed that the world had reached its definitive order, and that there was no other "choice" than the one between the Russian, Chinese and Cuban lies that their dishonest controversies weakly nourished. [Herbert] Marcuse, full of illusions, still claimed to demonstrate to us the disappearance of the proletariat, which was supposedly cheerfully dissolved into the bourgeoisie, and Henri Lefebvre, disillusioned, was already chattering on about "the end of history." By confessing so maladroitly that the reality of the period was all that they dreamed, they were only taking their poor dreams for reality. But after 1968, they had to come to sad terms with the stupidity from which they suffered. Marcuse became resigned to keeping quiet, and Lefebvre returned to the fold by speaking on behalf of the French Stalinists.

Today, when the time of disorder once again begins to disturb the sleep of the dominant classes, all the pathetic ideologues who are short on ideas have lost their audiences, but they have found unexpected jobs as defense attorneys for the old world. In Italy, where the crisis is the most serious, they have lost all restraint and – one step ahead of subversion – pop up here to hastily don the togas of the fatherland and appear over there like old cuckoo clocks to ceaselessly strike our ears with the same banalities about the defense of the republican order and the customary trivialities in favor of the democratic institutions with the same feigned and self-important conviction of the priests of a church that lacks a loyal congregation because faith is lacking in the miracles that they promise: namely, that history will stop, as if by enchantment, due to their magical formulae.

Every time that they show up on television or in one of the newspapers that imprudently invites us to appreciate the delights of the democracy that – fuck! – was born from the Resistance, just as they were born from the estimable cunts of their mothers, people like Valiani, Amendola, Asor Rosa, Moravia, Bobbio, Bocca et. al demonstrate that they do not want to understand that the violent and contradictory events that feed the columns of the newspapers only prove that *their era is over* and that a new world is here. These old caryatids that hope to support the desecrated and crumbling temple of the dominant lies and abuses for a little while longer – these extremists of consensus and fanatics of legality – do not know that their laws do not control the future or that, before judging the new men, one should judge the old laws. And, furthermore, the "democracy" and "liberty" with which these gentlemen gargle and assault our ears are, for them, *what colors are to someone who is blind from birth.* The proof of this is simple. If they knew the real meanings of these words, they would not make use of them so casually when speaking of our miserable republic. But when *real democracy* imposes itself – that is to say, when all decision-making and executive powers belong to the revolutionary workers' councils, from which every delegate is revocable by the base at every instant – well! then we will see all these gentlemen who today speak nonsense about democracy either combat it or, more probably, *flee from it,* as is their habit. But faced with the peremptory and insolent appeals with which these gentlemen gratify us today, the young proletarians are obligated to conclude that, if these respected hoaxers are solid in their courageous defense of all the current lies and abuses, it isn't by chance, but is in fact because they collect large payments for doing so. How many millions does honest Leo Valiani[11] receive each month, or each week, to write what he writes? And what would he write if he had the salary and life of a worker? And Bocca? And all the others?

Lichtenberg said that he didn't know a man in the world who, having transformed himself into a blackguard for a thousand *thalers,*[12] would not prefer to remain honest for half that sum.

Disappear, grotesque masquerade, charlatans of incurable diseases: you fear too many things to be feared and respect too many things to be respected! You judge everything wrongly, while people are beginning to judge you rightly. Do you not know that half of the country is laughing at you and the other half ignores you? You should at least know that, faced with the tragic-comic farce that constitutes your very existence, the court martial of our critique will soon celebrate its saturnalia! And one should not reproach me for having recourse to invective. Ever since Dante, all those who have regarded the powerful and their servants with disabused eyes have always been *constrained* to have recourse to invective. Because it is not enough to judge the acts and speeches of men; one must also *judge men according to their acts and speeches.*

Until now, the entirety of the country has remained spectators of its governmental ministers and all those who deceive it and speak in its name. Today the country must begin to judge them and to render unto Caesar what is Caesar's: twenty-three thrusts of the sword.

---

[11] Leo Valiani (1909-1999) was an anti-fascist militant and organizer of the Resistance during World War II. In the 1970s, he wrote for *L'Espresso,* a leftist news magazine. In the words of an obituary written by Philip Willan and published in *The Guardian* on 21 September 1999, "In the 1970s and 80s, his was one of the firmest voices to denounce the violence of politically motivated terrorists. He criticized the laxness of Italy's penal laws and even went so far as to argue for a restoration of the death penalty. He insisted that the state had to be defended against all sources of violence, whether the originators were fascist thugs, terrorists or mafiosi."

[12] Silver coins.

Gianfranco Sanguinetti

* * *

In the eras in which intelligence reigns, one can judge men and women according to the use they make of it; in the centuries of decadence, which nevertheless include skillful people, one must judge men and women according to their interests and merits; and in those periods when extreme mediocrity collides with great difficulties, which is the case today, one must consider the general conditions in which men and women live, the pretentions, fears and particular interests of those in power, and make our judgments based upon this mix. Thus, if today we witness the edifying spectacle that is offered to us daily by all the defense attorneys for the old world, who take the floor with ardor and haste to ejaculate their pleas in turn or all at the same time, this is because they fear that each time might be their last and because they all feel – with confusion, but not without good reason – that the tribunal of history is at the point of handing down a sentence that only comes too late. And if in their vain oratories these mercenary defenders of all the abuses sometimes seem reckless, this is only because, when fear has passed a certain limit, courage and cowardice can temporarily produce the same effects.

If the politicians and intellectuals are temporarily agitated by the word *courage,* it is principally to ask each other *what it is* exactly. And if, after the clamoring has ceased, they have not been able to give a response, we need not look very far to find the reason. As a general rule, men and women always speak the most of what they lack the most, and particularly in the situations in which they have the greatest need of it. Thus, while a poor person speaks of money, Franco Rodano speaks of courage. Lama, Moravia, Arpino, Calvino, Vasco Pratolini, Elio Petri and a hundred others try to outdo each other in their discussions of it. It isn't only Antonello Trombadori who has spoken of it – and, at least on this occasion, he showed himself to be reckless by speaking of rope in the house of the hanged man. And almost all of them have spoken of courage to accuse Montale and Sciascia[13] of cowardice, even though these last two have at least had the *minimum courage* to express publicly the disinterest and disgust that this State inspires in them, while the Stalinist Amendola[14]

---

[13] Leonardo Sciascia (1921-1989) was a writer who later became a politician affiliated with the Italian Communist Party, from which he resigned in 1977. In the years that followed, he was elected to the European Parliament and devoted himself to investigating the kidnapping of Aldo Moro.

fears to see the State collapse before the Christian Democrats have been able to share it with him.

All this demonstrates that one can say of courage what Marx said of consciousness. It is certainly not the courage of men and women that determines their social condition, but, on the contrary, it is their social condition that determines their courage and cowardice; and it suffices to consider the provisional character and current fragility of the social positions that these usurpers occupy in a society as uncertainly divided into classes as ours is to be sufficiently informed of their alleged "courage." Moreover, it goes without saying that no one has asked them to be [truly] courageous.

Cowardice has always existed, even if every age has not had the chance to see it in power. In our age, cowardice would like to be in the majority, but *it already is* the majority of the government, it has its heroes, and it publicly awards itself the dignities and honors that were, in other times, reserved for courage. All of the political-intellectual controversies about courage have only made obvious the profound cowardice of all those who have participated in them, since, if one cannot give oneself courage, one cannot take away one's cowardice. Indeed, one *cannot even hide it,* because I have never known a coward who at least has had the simple courage to recognize his cowardice and thus hide it better.

These periodic, weak and boring "polemics," which constitute the principal pastime of all the *eunuchs of power* – that is to say, the intellectuals – once again demonstrate the incurable weakness of those who participate in them. The weapons of their "critiques" do not fire because they are, as Camoens would say, "covered with the rust" of the social peace that they have enjoyed for too long, but only until recently. And we know that weakness is perhaps the only fault that one cannot correct, precisely because its effects are unimaginable and even more prodigious than the effects of the most vivid passions.

These courageous gentlemen defend the archaism of society's institutions only to avoid the misfortune of having to defend themselves. Nevertheless, they no longer even know how to get these institutions *to function,* and their archaic nature cannot get these men respected or venerated. Quite the contrary, these institutions discredit themselves every day, and they age even more rapidly than their coryphées. And, as their decadence becomes ever more obvious, they inspire a contempt that becomes

---

[14] Giorgio Amendola (1907-1980), a member of the Italian Communist Party, was a writer and member of the Constituent Assembly.

so universal that they are less and less in a position to do harm. Thus, the political world has fallen into a disastrous imbecility at the very moment when society as a whole has become *more intelligent.* Today, this imbecility and this intelligence harm power to an equal extent, and power finds itself constantly eaten away from the inside and attacked from without.

The social war that is coming has already put into motion all the individuals and all the classes of society because, by putting the interests of everyone back into play, it confers on everyone an interest in the [outcome of the] battle and it calls upon each person to choose his or her camp: on the one side, all those who fear a war *that they can no longer prevent* (the capitalists and the bureaucrats of the so-called Communist Party); and, on the other side, all those who have no power over their own lives and *know it.*

* * *

In the following chapters, I will write against the order of existing things, but I will do so *in relative disorder.* If I were to deal with this order in an orderly way, I would be according too much honor to my subject, because I want to show that *it is incapable of it.* As Saint-Just already said, "the present order is disorder put into laws." Before concluding this preface, there is hardly no need of saying that *Remedy for Everything* does not want nor can it be a remedy *for everyone.* Indeed, it proposes *to harm* many and hopes to be useful to an even larger number. The utility of such a pamphlet will thus be measurable *according to the damage* it is capable of causing, directly or indirectly, immediately or in a little while, to the owners of alienation, because everything that is harmful [to them] is not useless for this purpose. Only that which is useless is harmful [to us]. I hope to be *clear,* but if someone persists in not understanding, I will preoccupy myself less with this than he or she will be. It is said that this era can no longer be unconcerned about what it produces and, if it produces certain books, this means that it also produces *those who know how to read them.*

The owners of this world, as well as its salaried "critics," will be exasperated and vexed by seeing that only their most irreducible enemies are in a position to really understand it, and the dominant class will see with a justified inquietude that its real problems are only exposed by those who work at its subversion. Our government ministers and all the politicians will be justly disturbed by having to read our writings to finally be able to contemplate themselves with realism, but in the perspective of the destruction of all their powers. The heads of the bourgeoisie's secret services

– for a dozen years predisposed to provocations, assassinations and State terrorism – will justly be made furious by seeing their maneuvers constantly unmasked by the very people against whom these crimes were conceived, and even the death of Moro will finally appear in its true and sinister light.[15] The great decomposing bourgeoisie certainly will not want to pardon me, either for this pamphlet or any of the rest, and some among them – like Indro Montanelli, who has already cried about it for the last two years or so – wants to accuse me of being a *traitor to my class*, because I have turned all of my weapons against the aforementioned high bourgeoisie, from which I have come. Well, I am honored to receive such an accusation, because there is no humiliation (nor anything else) that this bourgeoisie has not amply merited, and the working class, which has been subjected to the largest number of class betrayals on the part of its alleged representatives, will have reason to congratulate itself because *for once*[16] their class adversary has been struck by the same fate.

Thus, *Remedy for Everything* will also be a *settling of accounts* with the entire underworld that the dominant class democratically imposes on the dominated classes, and also a settling of accounts with this or that *precise* person who has, until now, abused with too much impunity the patience of the exploited classes or, rather, the silence to which they are reduced. As in Hell, here one will find various graves and many damned souls – bourgeois and Stalinist, professional liars and labor-union bureaucrats, politicians and intellectuals, among others – with the result that, at the end, I, too, will be able to say to the reader:

> You can now judge these people,
> Whom I have accused here, and their faults,
> Which are the cause of all your misfortunes.[17]

---

[15] Aldo Moro (1916-1978), while still the Prime Minister of Italy, was kidnapped and murdered in an attempt to stop the "historic compromise," which he had long championed.

[16] Latin in original.

[17] Dante Alighieri, *Paradiso*, VI, 97-99, translated by C. H. Sisson (Oxford University Press, 1993).

Gianfranco Sanguinetti

Preface to the French Edition of *On Terrorism and the State*

If many books on terrorism are published in Italy, few of them are as little read as this one and none are as ignored by the press. Published at the end of April 1979, distributed slowly in a limited number of bookstores, *Del Terrorismo e dello Stato* was out of print by the beginning of the summer and has not been reprinted in Italy because of several difficulties created for me by a stupid and crude judicial-police persecution to which I will return. It is more interesting to ask oneself here, at the beginning, about the reasons for the quasi-complete silence that has surrounded a book that deals with a subject that is spoken about every day, but always in the same mendacious way, on the front pages of all the Italian newspapers as well as on the State-sponsored radio and television stations. Apparently my book was discussed in an *ad hoc* program that preceded a regularly scheduled installment of a television news-magazine, but, as several people have reported to me, this was only done so that a motley collection of experts on terrorism, brought together for the occasion, could say that the theses of my book "are not convincing." The most curious fact is that neither the television news-magazine nor the newspaper that wrote about it have ever dared to evoke these famous "theses" on Italian terrorism, which they nevertheless hastened to describe as "unconvincing." On the contrary, do they fear *that they are convincing* and is this why they keep silent about them with so much zeal? Do they fear that my arguments are in fact considered by [some] people to be more persuasive than their maladroit fantasies about terrorism, since all these reporters have made it their duty to make no allusions to them? If so, why so many precautions? What the devil is written in a book that is so scandalous that its existence is kept secret by the very people who are believed to have the obligation to speak about terrorism? Does *On Terrorism and the State* contain State secrets?

Well, yes: *this book contains State secrets.* Is not the principal secret of the Italian State the fact that its own secret services[18] have organized and pulled the strings of terrorism? And this is precisely the very thing that is amply demonstrated in *On Terrorism and the State.*

* * *

That which is *not convincing* is not my argumentation, but the self-

---

[18] The agencies in charge of gathering intelligence and covert operations.

contradictory behavior of the State and its loyal servants with respect to my book. On the one hand, they speak of it to say nothing about it, that is, when they aren't trying to have the Italians think that what I say isn't "convincing." On the other hand, several days after the televised "review," the political police and a judge who is known for the unfortunate zeal that he employs in trying to render probable all of the official lies on the subject of terrorism began a complex and solemn judicial-police persecution of me. Thus, should I think that I committed the crime of not having been "convincing"? If our [Criminal] Code makes provisions for such an offense, all the prisons of Europe would not be enough to hold our politicians, journalists, judges, police officers, union leaders, industrialists and priests. No, this was not why I was persecuted. I was in fact persecuted because I was *too convincing* when I accused the State of the crimes that this same State then sought revenge for, but, as we will see, with the embarrassed clumsiness proper to guilty parties who want to pass themselves off as innocent. The men who govern this State are, as one knows, the same ones who governed during the massacre at the Piazza Fontana[19] and, so as to not be accused of perpetrating it, they have been continuously obligated to accuse other men of their own crimes and all other crimes, as if they want to give a supplementary and practical confirmation of the theory of Madame de Stael,[20] according to whom "the life of any [political] party that commits a political crime is always linked to that crime, either to justify it or, *by virtue of its power,* to make that crime forgotten."

A series of disparate accusations – so crudely false and arbitrary that, one after the other, they collapsed without my attorneys having to intervene – were made against me during these past six months and, according to the whims of those who imagined them, they ran from smuggling to terrorism, and, naturally, included the possession of weapons and subversive association.

Of all these accusations, which could have brought me 20 to 30 years in prison if one kept to the letter of the law or, on the contrary, [if they were pursued] would have covered in ridicule those who made them against me, there were two of them that could have found a basis in reality (if one were to take them seriously and in a certain manner), while the others were completely false and bizarre.

I have indeed been a smuggler, but an honorable one. Since 1967, have I

---

[19] 12 December 1969.

[20] Anne Louise Germaine de Stael-Holstein (1766-1817).

not smuggled in from France the driving ideas of modern revolution, that is to say, the ideas of the Situationist International? And I also admit that, judging from the conditions in which the Italian State has found itself since then, this smuggling of the French disease has not benefited it: the contagion has been more rapid and deeper here than elsewhere, and the illness has not been eradicable. Unfortunately for my accusers, under the terms of our Code, as well as the Helsinki Accord, the smuggling of ideas is not condemnable, and we know quite well that, when the Italian State is concerned with ideas, it is surely not to get them cleared through customs. The accusation of smuggling thus collapsed miserably, even if it desperately sought, but unsuccessfully, to camouflage itself behind other common law pretexts.

As for the accusation of subversive association, although I do not exactly know what "subversive association" means in the context of the old fascist Code that is still on the books, I recognize that it, too, could have a basis in reality, since I belonged – in broad daylight and not clandestinely – to the Situationist International until its dissolution, which took place in the faraway year of 1972. I find this *inquisitio post mortem* of the SI laughable. A judge concerned with fairness, in addition to investigating the SI, would also have to open investigations into Marx's Communist League and the International Association of Workers, and issue a warrant for the arrest of the descendants of all those who housed Bakunin during his stay in Italy.

The accusation of weapons possession was based on absolutely nothing, and it certainly wasn't better founded for having been brought against me several times, but always without success. Contrary to the nattering of President Pertini, it seems to me that the civil war still hasn't begun – the proof of this is that he is still President of this thing that resembles a republic – and thus it is useless to possess weapons. And, in any case, those who accuse me of possessing weapons must at least find them, or at least plant them on me, and neither has happened yet.

Adding the arbitrary to the most obtuse arrogance, the Republic's chief prosecutor claims that, "according to the contents of the documents of the Red Brigades, there exist close connections between the ideology of this group and that of the Situationist International, of which the aforementioned Sanguinetti is a representative." Beyond the fact that the Italian section of the SI has not existed since 1970 and that, as a result, I cannot be [one of] its "representative[s]," and beyond the fact that the SI never had an *ideology,* because it fought against all of them (including the ideology of armed struggle), one must note at least two things. First, it would be less unfruitful if the judges [in question] educated themselves before making accusations.

Second, it would be much easier to show "close connections" between the police-like ideology of the aforesaid prosecutor and that of the Red Brigades than between the ideology of the Red Brigade and situationist theory. And nothing in the world is more radically opposed to what I wrote about the Red Brigades than what this group says about itself, with the support of both the bourgeois and the bureaucratic press. Finally, I note – so as to not rely on arguments that are too easy – that is easy to buy the publications of the SI in Italy and that there are many people who know them, despite what the voice of this or that imprisoned Autonome[21] says. Furthermore, anyone who reads these publications can ascertain that in no case does there exist "close connections" between these writings and the documents of the ghostly Red Brigades, despite what that impertinent prosecutor claims.

* * *

In the same way, and at the same time that the authorities were conducting this maladroit persecution – which was filled with low blows but at least had the merit of being public and official, as were the incriminations, searches, surveillance and wiretaps connected to it – obscure and vile people who were by their behavior easily identifiable as cops, acting with fewer scruples but with no more success, operated in the shadows with the aim of provoking or intimidating me. Not being an intellectual, and not having the necessity of making a living from what I write, I have never claimed to receive better public recognition than this for what I myself publish, at my own risk and peril, at a time and in a country where no one dares to run the risk of saying to people *that which one doesn't want people to hear,* that is to say, the simple truth about terrorism and the rest.

For the benefit of readers who don't live in Italy, and to give Italy the publicity that it merits, I will add that several travelers have been stopped at the border by the Italian police, taken by force to a large city and interrogated at length for the sole reason that they possessed a copy of this book; that the magistracy has opened an investigation into those who have distributed it; and that the DIGOS,[22] without even having obtained a proper seizure order, has arbitrarily seized whatever copies it has been able to find.

Thus, there is no longer any doubt, if there ever was one: *I have told the*

---

[21] A member of *Autonomia Operaia* ("Workers Autonomy").

[22] The *Divisione Investigazioni Generali e Operazioni Speciali* (DIGOS), which is officially tasked with fighting organized crime, terrorism and capital offenses.

truth. And, thanks to the harm one wants to do to me, I know that my work is good, and I certainly would not have provoked such hatred if no one had heard me. In fact, among the people who have read what I have written, and who are of various ages, conditions and opinions, many have approved, few have doubted, and none have refuted me.

* * *

Since the first edition of this book, many events have taken place, none of which have necessitated the least modification (in either the entirety or the details) of its arguments or conclusions. Indeed, these events have only confirmed them. We have witnessed the elimination of Alessandrini, a judge who was becoming troublesome, first for taking apart the rigged trial [of those accused of the bombing] of the Piazza Fontana, and then – several hours before being killed, officially by [Leftist] subversives – having questioned an ex-chief of the S.I.D.[23] about his false testimony and the false testimony given by his superior officers, Andreotti and Rumor,[24] at that rigged trial. Then we saw a disciple of Moro, the Honorable Mattarella, President of the Sicily, meet the same end as his master, and *for the same reason*, on the eve of his formation of the first regional government of "compromise" between the Christian Democratic Party and the Italian Communist Party [ICP]. On diverse occasions, we have also seen several police officers suddenly get iced so as to heat up support for and nullify all opposition to the *villainous laws*[25] that surpass and invalidate the old fascist laws (thought to be too tolerant), as well as the republican Constitution. But the most important of all the novelties that appeared during this past year was certainly the fact that the ICP – seeing its prospects for active and immediate participation in the government evaporate with the death of Moro – adopted a fallback position, which was to make a warhorse of its active participation in the spectacle of terrorism and its repression. This is clearly the principal novelty to appear after the publication of the first edition of this book, and it merits a few remarks because it once again demonstrates that, not only do the Stalinists know that it is power that perpetrates terrorism, but also that those who wish to be in power in Italy

---

[23] The *Servizio Informazioni Difesa* (DIS), a defense intelligence agency.
[24] At the time of the investigation, Giulio Andreotti (born 1919) was the Minister of Defense and Mariano Rumor (1915-1990) was the Minister of the Interior.
[25] French in original Italian version.

today must demonstrate that they know how to perpetrate terrorism. And this is so true that even a former government minister (a Socialist) recently declared in an interview that, "in Italy, one makes policy with terrorism."

Until 7 April 1979,[26] the ICP contented itself with issuing stupid, ritualized appeals against terrorism, which it defined by feigning to believe all the official versions of the attacks, thus proving its good will to the Christian Democrats and its bad conscience to everyone else. But from that day on, the Stalinists, through the intermediary of the magistracy, began to put to good use their vast and rich fifty-year-long experience with the discovery of fake suspects, the staging of rigged trials, the production of false testimony and prefabricated proofs.

Since their double goal was to show their merits to the Christian Democrats and to get rid of a limited but embarrassing force (because it was situated to their left and insulted them), the Stalinists found among the Autonomes the guilty parties for 10 years' of assassinations, massacres and [other acts of] terrorism. There wasn't a crime committed in the 1970s that wasn't committed by this or that Autonome. From unsolved murders to the Moro affair, from mysterious kidnappings to the thefts of works of art and racehorses, every crime was solved, suddenly and as if by magic; every offense found its guilty party and every guilty party found his or her compensation in a prison sentence. To obtain such a harmonious settlement of the trials of the past decade, *the genius for harmony and invention*[27] of a simple Stalinist magistrate was certainly not enough. The entire organization (both its hidden and public parts) of the ICP was mobilized with the goal of proving that *Autonomia was* the armed struggle. As if by chance, the only Autonome leader to remain at liberty, the naïve Pifano, was very quickly caught with his hand in a sack that contained two non-functioning Lance missiles, furnished to the aforementioned Pifano by the FPLP,[28] a Stalinist Palestinian organization that was, according to General Miceli himself,[29] notoriously linked by reciprocal

---

[26] The day on which the Italian authorities began arresting dozens of alleged subversives, including Antonio Negri and Franco Piperno.

[27] Here Sanguinetti has détourned the title of a well-known series of violin concertos by Vivaldi: *Il Cimento dell'armonia e dell'inventione,* which is commonly translated as "The Trial of Harmony and Invention."

[28] *Front Populaire de Libération de la Palestine* ("Popular Front for the Liberation of Palestine") founded in 1967.

[29] Vito Miceli (1916-1990) was the head of the S.I.D. until his arrest in 1974 for assisting in a failed right-wing coup d'état.

recognition to the Italian secret services. Thus, if until that moment the links between *Autonomia* and terrorism could not be demonstrated, the zealous Pecchioli[30] was, several hours later, able to declare to Parliament that, faced with such an eloquent fact, no one had any right to doubt that the Autonomes constituted the strategic leadership of terrorism, which was what the Stalinist magistrate Calogero[31] had already maintained, but without any proof. The poor Autonomes, who for their part have never understood terrorism or revolution, have thus ended up – such coveted prey – in the carnivorous jaws of the Stalinists and the magistracy, without even understanding why or how. One must hope that, where their self-instruction is concerned, they make better use of their time in prison than they did when they were free.

In both their ingenuity and their crudeness, the Stalinists' admirable methods of accusation are not at all original, but closely match the ones used in the famous Moscow "show trials" of the 1930s. The only difference is that the arrested Autonomes have still not been declared guilty of all the crimes that have been committed. The incongruity of these most-recent juridical procedures must not be held against the Stalinists. No doubt they would quickly disappear if the Stalinists had control of the police forces and could make use of their tested and infallible system during the interrogation phase.[32]

For the secret services and the ringleaders of the Christian Democrats, who have suffered so many judicial humiliations these past few years – certainly not due to the honesty of the judges, but their own lack of ability – the great trials of the Autonomes, which have been so skillfully mounted, have opened up unexpected perspectives and new fields of action. Indeed, ever since 7 April 1979, the spectacle of terrorism has made immense progress and, if, until then, the secret services had been compelled to go too far, now that the Stalinists have shown themselves to be such skillful and unconditional allies, there is reason to believe that, like Ulysses, they will make themselves "wings for the foolish flight, always gaining a little on the port side."[33]

---

[30] Ugo Pecchioli (1925-1996), a member of the Italian Communist Party.

[31] Pietro Calogero (born 1939), the public prosecutor in Padua.

[32] That is, just throw the accused out the window and kill him that way, as was done on 15 December 1969 with Giuseppe Pinelli, allegedly one of the perpetrators of the bombing at the Piazza Fontana. Note that in Richard N. Gardner's book *Mission Italy: On the Front Lines of the Cold War*, published in 2005, Eugenio Scalfari is quoted as saying (in 1978) that "terrorism will be conquered in Italy only when Berlinguer is in charge of the police."

Acting in this way, the bureaucrats of the ICP have done nothing other than what they are capable of doing and incapable of not doing when they find themselves within reach of power. They know perfectly well that, this time more than ever, they have every reason to be dishonest, because it is in the current period that their historic enterprise is being played out, and it is natural that they seek to put into play all of their forces, since their entire fortunes are at stake.[34] They have a supplementary reason to immediately show all of their historical dishonesty. They are certainly not ignorant of the fact that it is uniquely because of their dishonesty, and not for their very well-hidden virtue, that the bourgeoisie can now employ them in their service. And, more precisely, the Stalinists know that they must continuously invent and discover conspiracies against bourgeois democracy, either to feign to love it better or to show to the world all the dangers that it runs without their help.

If the ICP behaves this way in public, it surely acts with the same contemptible baseness in its "private life" at the factories, that is, by indicating to the bosses the identities of the "terrorist" workers (those who do not want to submit and practice absenteeism, that is to say, those workers who *struggle*), so that they can be fired and denounced in the name of the justice of work.

Contrary to the hopes of subtle Berlinguer,[35] the bosses and the best-advised men in the Christian Democratic Party have concluded that the more the ICP shows itself to be useful without being a part of the government, the more useless it would be to bring them into it, with the result that all that the Stalinists do (and by all possible means) to attain power is in fact what keeps them out of it, which further alienates them from the electoral sympathies and illusions that they had once garnered. But this is the drama of the Stalinists and it doesn't concern us, at least insofar as they have not become malicious enough to return to the practice of their preferred art, which is political crime. Until then, we must note about what immediately concerns us that bourgeois terrorism and Stalinist terrorism, both of which seek the same goal, reveal themselves to be what they have always been, and give the working classes an excellent occasion to recognize and combat *all* of their enemies, both bureaucrats and bourgeois.

---

[33] Dante, *Inferno*, XXVI, 125-126. Note that the port side of a ship is its *left* side.

[34] Machiavelli, *Discourses on Livy*, Book I, Chapter 23.

[35] Enrico Berlinguer (1922-1984), the Secretary General of the Italian Communist Party. Note that Chapter IV of Sanguinetti's *Remedy to Everything* was devoted to "Invective Against Enrico Berlinguer."

* * *

The active servility with which the entire Leftist intelligentsia at first tolerated, then adopted, the official accusatory theses about terrorism and the Autonomes' connections to it could seem properly stupefying to someone who doesn't know that this intelligentsia has always acted in this fashion every time that it has had the opportunity to act otherwise. The governmental-Stalinist version of the facts was accepted point by point and thus publicized without the least respect for historical truth or so-called "intellectual dignity." Furthermore, it is notorious that, for the last half-century, the role of Italian intellectuals, who are Stalinist for the most part, has been irreplaceable in the diffusion of all the lies on the subjects of socialism and revolution. Today, since they can no longer lie about Soviet, Chinese and Cuban "socialism," they have been reduced to the spreading of limitless lies about bourgeois democracy. To safeguard it, they have willingly made every sacrifice and have even safeguarded it without making any sacrifices at all. Thus, the recent government decrees concerning [police] custody, the penalties for crimes of terrorism and "possession of subversive documents" were approved without protest and in homage to the fetish of democratic guarantees. The new provisions on preventive detention will permit the State to keep an accused person in prison *for 12 years without trial.* From now on, the Italian magistracy, for which the existence of courtesans has never been a State secret and no longer has to be proved, does not have to bother with demonstrating the guilt of whomever is *de facto* condemned to 12 years in prison, and this is just the beginning. From now on, the accusation coincides with the condemnation, and the fiction of democratic liberty in Italy has ended *even as a fiction.* Italy is a democratic republic founded on the exploitation of work[36] and executive orders.

In a passage in *The Phenomenology of Mind* that is little known by our intellectuals and that relates to governmental terrorism, Hegel says,

> By no manner of means, therefore, can it [the government] exhibit itself as anything but *a faction.* The victorious faction is only called the government, and in the fact that it is a faction lies the direct necessity of its overthrow, and its being [the] government makes it, conversely, into a faction and hence guilty (...) Being suspected,

---

[36] Cf. Article I of the Italian Constitution.

therefore, takes the place, or has the significance and effect, of being guilty.[37]

When the arbitrary *no longer fears* to appear as what it has always been, and when being guilty or innocent no longer has *any importance,* since the condemnation becomes *the only certitude,* then anyone who combats the arbitrary no longer has to fear *being guilty.* He or she is condemned for being condemned as much as for committing an honorable crime. Thus, we cannot let ourselves be governed innocently. And so, waiting to destroy all the prisons, let us give the enemy good reasons to fill them, certainly not by falling into the well-set trap of terrorism, but rather by fighting openly and by all means all those who today make use of it and practice it: government ministers, politicians, bosses and police officers.

In our times, intellectual Jesuitism calls democracy "the arbitrary," freedom "the freedom to lie," and testimony "systematic and obligatory informing." "*Thus were informers, a race of men brought forward for the destruction of the public and never sufficiently restrained by pains or penalties, allured and invited to action by rewards,*" Tacitus said,[38] though he – unlike our intellectuals – confessed that he preferred the dangers of freedom to the tranquility of slavery. These same intellectuals, after having debated courage back and forth, and up and down, proudly concluded that today one must have the courage to be a coward. The reasoning in fashion these days is simple: if one loves democracy, one must defend it; to defend it, one must combat its enemies; to combat the enemies of democracy, no sacrifice is too great; the nobility of the goal justifies every means; no democracy for the enemies of democracy! That which was *essentially* not a democracy has now *visibly* ceased to be one.

And who are [supposedly] the real enemies of democracy? All those who *objectively* put it in danger by propagating ideas that are incompatible with it, and all those who *objectively* support its enemies by not supporting the State. In sum, the enemies of *this* "democracy" are all those who practice democracy.

In 1924, had this "democracy" – so sincere, so quick to pretend to be the contrary of what it really is – been in power, instead of Mussolini, one could be certain that the means would have existed to accuse the Leftists of the

---

[37] G. W. F. Hegel, "Absolute Freedom and Terror," *The Phenomenology of Mind,* translated from the German by J. B. Baillie.
[38] *The Annals,* Book IV, Section 30. Latin in original.

assassination of Matteotti,[39] just as one accused the Leftists of assassinating Moro in 1978. But as Mussolini had less need of lies than the current State, he would not have needed to employ intellectuals such as Leo Valiani[40] to speak to us of State crimes with the same admiration that one speaks of the virtues of Cato [the Younger].

I know quite well that the Italian intelligentsia has a number of reasons to be fearful and dishonest; I even know its self-justifying arguments by heart; and I would never dream of refusing it the freedom to be contemptible. What I find tedious is the fact that intellectuals constantly intervene with respect to terrorism in the newspapers and weekly magazines, as if an obscure force pushes them to publish the proofs of their unlimited baseness, and as if it were still necessary to convince anyone of it, whereas they might be better off if they confined these proofs to their books, so that their baseness is not known to either posterity or their contemporaries.

For example, no one among the great thinkers in matters of terrorism has yet formulated the simplest of arguments on the subject. If the ghostly Red Brigades [RBs] were a spontaneous grouping of subversives, as one says that they are, and if Negri and Piperno were the leaders of the RBs, as one pretends that they are, then why would that clever group allow the State to imprison its leaders, who deny that this is what they are, and why would they not even seek to exonerate these men, if only to try to recuperate them afterwards? If, on the other hand, Negri and Piperno are not the leaders of the RBs, and are not even members of that organization, then the hypothetical subversives of the RBs would have even more reason to publicly clear them of these accusations. Indeed, they would have three good reasons for doing so: to not allow leaders to be falsely attributed to them without protest; to not be accused of allowing innocent people to be condemned in their places; and, finally, because – being protected by clandestinity – they run no risk of clearing the names of the people currently accused.

Since nothing of the sort has happened, we must conclude that the real leaders of the RBs have the same interest as our State does in having it believed that Negri and Piperno are their leaders. This novel convergence of interests between the State and the RBs has nothing fortuitous or extraordinary about it and can only stupefy the stupid people who have not realized that the RBs *are the State,* that is to say, one of its armed appendages.

---

[39] Giacomo Matteotti, a socialist leader who was murdered by the fascists.
[40] In April 1945, Valiani was one of the signers of the document that ordered the execution of the Italian dictator.

Therefore, even these few simple deductions, which, on their own, suffice to prove the enormity and fragility of the generalized lies about terrorism, are too bold to be formulated by our free thinkers, who are so free that they have come to the point of *no longer thinking.* On the contrary, they revel in trying to outdo each other with sub-Machiavellian and maladroit theories, such as the one that tries to prove that the dissolution of *Potere Operaio,*[41] which occurred six or seven years ago, was a diabolical simulation that allowed its leaders and militant members to better devote themselves to armed struggle. And this has been repeated for months without anyone perceiving that the hypothesis is absurd, and for the very good reason that it invokes. If *Potere Operaio* had truly been the cover for terrorist activity, why would its leaders deprive themselves of such a valuable legal front?

The truth is completely different and, as is customary, to find it one must reverse the shameless lie with which one would like to camouflage it. It certainly isn't *Potere Operaio* that feigned to dissolve itself to better devote itself to terrorism, but the famous S.I.D. that feigned to dissolve itself to have its past terrorism forgotten and to practice it better thereafter. Other salaried thinkers, from Scalfari to Bocca,[42] reason fraudulently when – admitting that the strategies of the RBs, among other things, aim at preventing the arrival of the ICP into power, which is what I have demonstrated – they believe that this aim results, not from the aversion that the Communists cause in certain sectors of Italian capitalism and its secret services, but from the aversion that the Soviet Stalinists feel for their Italian counterparts. Our thinkers who are paid by the week thus conclude that Moro was kidnapped with the support of the KGB and the Czech secret services. The Italian capitalists, the military men and the agents of the SISDE, SSMI, CESIS, DIGOS and UCIGOS,[43] as well as [the American President Jimmy] Carter, would be happy to see the ICP in Italy's government, but this is, unfortunately, not possible because the Russians and the KGB don't want it to happen. What bad luck! If the KGB was behind the Moro affair, then who or what is behind Bocca and Scalfari, those idiots? And is it possible that they have been raised to such heights by their own strengths?

---

[41] "Workers' Power."

[42] Eugenio Scalfari (born 1924) was the editor of *La Repubblica,* which he founded in 1976, and Giorgio Bocca (1920-2011) was the author of *Il Terrorismo Italiano* (1978).

[43] The various Italian agencies tasked with gathering intelligence and conducting covert operations.

In any event, this curious and stupid theory, which impetuous Pertini[44] hastened to make his, clearly serves to reassure the bad consciences of all those who want to believe that this State, because it is at war with terrorism, cannot be directing it, too.

For my part, I note with legitimate satisfaction that my book, which at first forced silence upon all those who are paid to speak, then obligated them to talk all day and commit themselves to an interminable series of outrageous remarks designed to support the opposite of the truths that, with this book, began to circulate freely in Italy.

In an extremely different sense, here one can mention Russia, because contemporary Italy and Russia under Stalin are perhaps the only States in the world that are *exclusively* maintained by the secret police. In Russia, "counter-revolutionaries" were discovered everywhere, and anyone opposed to Russia was accused of being one. In Italy today, "revolutionaries" are discovered everywhere, and every extra-parliamentarian, even the most timid ones, are open to this accusation. According to the judges and newspapers, Negri, Piperno, Scalzone and the others would be the leaders of the Italian revolution, its brains and its strategists. I have defended them here as innocent people, but I would never dream of defending them as revolutionaries, because they are neither guilty nor revolutionary. In reality, all of the Autonome leaders are only naïve politicians, and even as politicians they are imprudent failures. One has never seen [true] revolutionaries dine with magistrates, as Negri has done, nor converse over a meal with an ex-minister of Mancini's type, as Piperno has done. Neither man is a [true] revolutionary for a thousand other reasons that are so obvious that it is useless to recall them. The Italian revolution follows a completely different course and completely different ideas, and it deliberately passes over these leaders, these brains and these strategists, just as it passes over all those who understand nothing about terrorism, that is to say, the counter-revolution.

* * *

One knows the passion that the freest people (the ancient Greeks, for example) had for the enigma, which they considered to be the *Hic Rhodus, hic salta*[45] of wisdom. Confronted with an enigma, the wise person must know

---

[44] Sandro Pertini (1896-1990), then the President of Italy.

[45] This Latin expression (a translation of a line in Aesop's fable "The Boastful Athlete") literally means, "Here is Rhodes, jump here." In his preface to *The*

how to solve it *at the cost of his or her life.* Solving it was a struggle in which the one who could not do so could not expect any indulgence. If one believes in the legend reported by Heraclitus, as well as by Aristotle, Homer, who was the wisest of the Greeks, died from despair because he had not been able to solve an enigma. He who does not manage to solve an enigma is deceived by it; he who is deceived is not a wise man; he who is not wise dies, because the wise man is a warrior who must know how to defend himself or succumb, because it is only in battle that he can prove what he is.

A eminent Hellenist has observed that the formulation of an enigma "contains the distant origin of the dialectic, called upon to open up without interruption the enigmatic sphere, according to the structure of the *Agon* and the terminology itself."[46] Nietzsche had already said that the dialectic "is the new art-form of the Greek *Agon.*"[47]

Therefore, Italian terrorism is *the last enigma of the society of the spectacle* and only those who reason dialectically can solve it. It is because of the lack of dialectics that this enigma continues to deceive and cut down all the victims that this State liberally sacrifices at its own altar, because it is on this unresolved enigma that it provisionally maintains itself. Thus it is necessary and sufficient to solve this enigma, not only to put an end to terrorism, but also to provoke *the collapse of the Italian State.* Only those who have an interest in doing so will resolve this enigma *in practice.* But who has an interest in untangling the enigma of terrorism? Obviously no one, *except the proletariat,* because only the proletariat has the necessary urgency, the motivations, the strength and the abilities necessary to destroy the State that deceives and exploits it. The goal of all the provocations of the last 10 years and the subsequent pedagogical campaign to indoctrinate the masses was to mastermind people's thinking, to obligate them to think certain things. With terrorism, the State hurled a deadly challenge at the proletariat and its intelligence; and the Italian workers can only accept it and demonstrate that they are dialecticians, or passively accept their defeat. Today, all those who

---

Philosophy of Right, Hegel – in an apparent reference to the Rosicrucians – offered an altered translation: *Hier ist die Rose, hier tanze* ("Here is the Rose, dance here"). According to Marx, writing in *The 18th Brumaire of Louis Bonaparte,* "a situation is created which makes all turning back impossible, and the conditions themselves call out: Here is the rose, here dance!"

[46] Theodor Adorno, "Enigmaticalness, Truth Content, Metaphysics," *Aesthetic Theory* (1970).
[47] Frederick Nietzsche, *Twilight of the Idols* (1888).

speak of social revolution without denouncing and combating the terrorist counter-revolution have a corpse in their mouths.

Having reached the height of imposture, the State has never felt so assured of itself, but in this it is much more deceived that it believes it is, because the State manages to deceive fewer people than it hopes and even fewer than it needs to. But, more particularly, this discredited State is deceived when it believes itself to be believed always, that is to say, when it believes that the lies propagated about terrorism by all the sources of information suffice to corrupt the entire population for the simple reason that there are no other sources. The proletariat, which, as one knows, has no means of expressing itself freely, thus cannot even express its legitimate incredulity concerning the tragic-comic farce of terrorism, at least to shut up (once and for all) the sycophants who speak of terrorism in the manner that we have described, as well as their constituents, who are precisely the constituents of terrorism and the beneficiaries of exploitation.

This being said, never – not even in wartime – has the Italian State, having recourse to systematic brainwashing, been able to corrupt so many minds so cheaply.

In contemporary Italy, everything that is obviously false and only what is false finds a home, sells itself, is purchased and is a source of profits. The staging and propagation of the infection of terrorism is a colossal and profitable enterprise that ensures the employment of tens of thousands of journalists, cops, secret agents, judges, sociologists and specialists of all kinds. "Only the truth has no clients," Montesquieu said in less mendacious times, but today the truth *has no need of them*.

* * *

I hope that this preface will help readers outside of Italy understand the forces, interests and fears that, in barely 10 years, have made it *the country of the lie and the enigma* (to adopt the title of the famous book by Ciliga on Russia under Stalin).[48] On this peninsula – cradle of modern capitalism, headquarters of the Papacy, center of Christianity and Euro-Stalinism, and privileged place for the counter-revolutionary experimentation that stretches from the Counter-Reformation, through fascism, to the current enterprises of

---

[48] The title of the Italian translation of Anton Ciliga's *Ten Years in the Country of the Disconcerting Lie,* which was first published in 1940 under the title *The Russian Enigma.*

the secret services and the Stalinists – where the vestiges of past grandeur attract so many visitors from abroad, the putrid wastes of the decomposition of all that marked the past millennium now flow and the entire population is besmirched by the fetid and foul smell of Christianity, capitalism and Stalinism at their ultimate stages of infection, all of them supporting each other for one more moment in the face of the menacing imminence of the most modern revolution, all of them meeting here to put to work the most merciless and desperate repression, and to argue about the most efficient system to condemn history, which has condemned them.

But whatever the events that await us, the only certainty is that they will obligate the Italian proletariat to make its own the phrase by Lucius Junius Brutus: *"I swear that I will never let either this person or any other govern Rome."*[49]

January 1980

---

[49] Livy, *A History of Rome,* Book 1, Paragraph 59. Latin in original.

Gianfranco Sanguinetti

On Terrorism and the State

The wily shafts of state, those jugglers' tricks,
Which we call deep designs and politics,
(As in a theatre the ignorant fry,
Because the cords escape their eye,
Wonder to see the motions fly) (...)
Methinks, when you expose the scene,
Down the ill-organ'd engines fall;
Off fly the vizards, and discover all:
How plain I see through the deceit!
How shallow, and how gross, the cheat!
Look where the pulley's tied above! (...)
On what poor engines move
The thoughts of monarchs and designs of states!
What petty motives rule their fates! (...)
Away the frighten'd peasants fly,
Scared at the unheard-of prodigy (...)
Lo! it appears!
See how they tremble! how they quake!

Swift, "Ode to the Honorable Sir William Temple," 1689

All acts of terrorism, all the attacks that have struck and that strike the imagination of men and women, have been and are either *offensive* or *defensive* actions. Experience has long since shown that, if they are part of a strategic offensive, they are always doomed to failure. On the other hand, experience has also shown that, if they are part of a defensive strategy, such actions can hope for some success, which is nevertheless momentary and precarious. The attacks by the Palestinians and the Irish, for example, are acts of *offensive* terrorism, while the bombing of the Piazza Fontana and the kidnapping of Aldo Moro, for example, are *defensive* acts.

However, it is not only the strategy that differs depending on whether the act in question is an instance of offensive or defensive terrorism, but also *the strategists.* The desperate and those suffering from illusions have recourse to offensive terrorism, while it is *always* and *only* States that have recourse to defensive terrorism, either because they have been thrust into some serious social crisis, as the Italian State has been, or because they fear such a crisis, as does the German State.

The defensive terrorism of the States is practiced *directly* or *indirectly* by them, that is, with their own weapons or with those of others. If the States have recourse to *direct* terrorism, it is directed against their own populations, as was the case with the massacres at the Piazza Fontana, on the *Italicus* or at Brescia.[50] If, on the other hand, the States decide they must have recourse to *indirect* terrorism, such acts must appear to have been directed against them, as was the case in the Moro affair.

The attacks directly realized by detached units or by the unofficial services of the State[51] are not customarily claimed by anyone, but are imputed or attributed to this or that convenient "guilty party," such as Pinelli or Valpreda.[52] Experience has proved that this aspect is the weakest point of this type of terrorism and that determines the extreme fragility of the political usage one wants to make of it. The results of this same experience show that the strategists of the State's unofficial services seek to give their own acts much greater credibility or at least less improbability, either by directly claiming them in the name of the initials of this or that ghostly group, or even by getting them claimed by an existing clandestine group, whose militants are apparently or believe themselves to be strangers to the designs of the State apparatus.

All the secret terrorist groups are organized and directed by a clandestine hierarchy that is composed of the militants of clandestinity themselves, who perfectly reflect the division of labor and the roles proper to

---

[50] The bombing of a bank at the Piazza Fontana in Milan took place on 12 December 1969. The bombing of the *Italicus Express,* a train operated by the *Ferrovie dello Stato* ("State Railway"), took place on 4 August 1974 and was at first "claimed" by *Ordine Nero* ("Black Order"), a neo-fascist group. The bombing at Brescia, Italy, took place at the Piazza della Loggia on 28 May 1974, during an anti-fascist protest.

[51] The agencies in charge of gathering intelligence and conducting covert operations.

[52] Giuseppe "Pino" Pinelli (1928-1969) was an Italian railway worker and anarchist activist. Accused of perpetrating the attack at the Piazza Fontana, he was murdered on 15 December 1969 by the Italian police, who forced him out of a fourth floor window. Pietro Valpreda (1933-2002) was an anarchist and writer who was accused and convicted of perpetrating the attack at the Piazza Fontana.

the current social organization: those on high decide on what is to be done and those below execute orders. Ideology and military discipline protect the true summit from all the risks and the rank-and-file from all suspicions. Any secret service can invent for itself a set of "revolutionary" initials and carry out a certain number of attacks for which the press will make good publicity and from which the secret service in question will find it easy to form a small group of naïve militants, whom it can direct with the greatest ease. But in case a small terrorist group spontaneously constitutes itself, there is nothing easier in the world for the detached units of the State to do than infiltrate it and then, thanks to the means at their disposal and the extreme freedom of maneuvering that they enjoy, to substitute themselves for it, either by well-chosen arrests made at opportune moments or by the assassination of the original leaders, which, as a general rule, takes place during an armed conflict with the "forces of order," informed in advance of such an encounter by the infiltrated agents.

From that moment on, the unofficial services of the State can dispose as they please of a perfectly effective organization, composed of naïve or fanatical militants who only ask to be led. The small original terrorist group, born from the illusions of its militants concerning the possibilities of launching an effective strategic offensive, changes strategists and becomes nothing other than a *defensive* appendage of the State, which maneuvers it with the greatest agility and assurance, according to its own necessities of the moment or those that it *believes* are its own necessities.

From the [bombing of the] Piazza Fontana to the kidnapping of Moro, the only things that have changed are the contingent objectives that this defensive terrorism has achieved, but *the goal* of the defensive *can never change.* And the goal from 12 December 1969 to 16 March 1978, and today, as well, has in fact remained the same: to make the entire population, which had not supported the State or had been struggling against it, believe that it *at least has an enemy in common* with the State and that the State will defend the population on the condition that no one questions it. The population, which is generally hostile to terrorism, and not without reason, must then agree that, *at least in this instance,* it *needs* the State, to which it must delegate the most extensive powers so that the State can vigorously confront the arduous task of the common defense against an enemy that is obscure, mysterious, perfidious, merciless and, in a word, *illusory.* Faced with a terrorism that is always presented as *the absolute evil,* evil in itself and by itself, all the other evils, which are much more real, become secondary and must even be forgotten. Because the struggle against terrorism [perfectly] coincides with *the common interest,* it is already *the general good,* and the State that generously leads that struggle is the good itself and by itself. Without the cruelty of the devil, the infinite kindness of God cannot appear and be properly appreciated.

The State, extremely weakened by all the attacks it has suffered every day for 10 years – attacks on its economy made by the proletariat, on the one hand, and attacks on its power and prestige made by the ineptitude of its managers, on the other –, can thus silence both them by solemnly tasking itself with staging the spectacle of the collective and sacrosanct defense [of all] against the monster of terrorism and, in the name of this pious mission, it can take from all of its subjects a supplementary portion of their already limited freedom and thus reinforce the police-related control of the entire population. "We are at war," and war against an enemy that is so powerful that any other discord or conflict is an act of sabotage or desertion. It is only to protest against terrorism that one has the right to the recourse of the general strike. Terrorism and "emergency," a state of emergency and perpetual "vigilance," become the only problems, at least the only ones with which it is permitted and necessary for people to be occupied. All the rest doesn't exist or becomes forgotten, and in any case is shut up, banished, repressed into the social unconscious because of the seriousness of the question of "public order." And confronted with the universal duty of its defense, everyone is invited to become an informer, to be base and to become fearful. For the first time in history, cowardice becomes a sublime quality, fear is always justified, and the only form of "courage" that is not contemptible is the one that approves and supports all the lies, abuses and infamies of the State. Since the current crisis doesn't spare any country in the world, there are no geographical boundaries between peace, war, freedom or truth. These borders pass through every country, and each State arms itself and declares war on the truth.

Someone doesn't believe in the hidden power of the terrorists? Well then, he or she must change his or her opinion when confronted with cleverly filmed images that show three German terrorists at the moment of boarding a helicopter, and they are so powerful that they even manage to escape from the German secret services that are better at filming their prey that catching them.

Someone doesn't believe that one or two hundred terrorists are in the position to deal a deathblow to our institutions? Well then, he or she will see that five or six of them are able to abduct Moro and his escort in a few minutes and will thus [have to] admit that the danger to those institutions (so loved by more than 50 million Italians) is real and terrible. Perhaps someone still believes otherwise? He is an accomplice of the terrorists! Everyone will then agree that the State cannot go down without defending itself and, whatever the costs, this defense is a sacred and imperative duty for everyone. And this would be the case because the republic is public, the State is for everyone, everyone is the State, and the State is everyone, because everyone enjoys its advantages, which are equally shared. Is that not democracy? And this is why the People are sovereign, but watch out if they do not defend democracy!

Are you convinced? Or do you, poor citizens in the mood for critique, still believe – in the wake of the Moro affair – that it is the State that has launched such attacks, such as the one at the Piazza Fontana? Vile suspicion! The dignity of the State's institutions is sullied by it. Zaccagnini[53] is crying: look at this photograph. Cossiga[54] is crying, too: look at this television news-magazine, and once and for all stop making accusations against all those who do not hesitate to sacrifice the life of another person in the name of the defense of our very democratic institutions! Or perhaps, poor citizens, you still believe that we, the government ministers, generals, and secret agents of "anti-terrorism" – to speak ironically – that we, in particular, would be disposed to sacrifice Aldo Moro, that remarkable statesman of elevated sentiments, that example of moral rectitude, our friend, leader, protector and, when necessary, our defender?[55]

That is precisely what one would not want to be thought by each good citizen (who never doubts, always votes, pays up if he isn't rich and, in any case, keeps his mouth shut). Suspicions about the State's role in the massacre at the Piazza Fontana are permitted, because the victims were [merely] ordinary citizens, but one would surely not want the State to be suspect when the victim is its most prestigious representative! Kennedy?! That kind of thing is a thing of the past.

This was precisely why the agony of Moro had to last for such a long time, so that each person, at his or her leisure, had plenty of opportunity to follow the spectacle of the kidnapping and the feigned discussion about the negotiations by reading the pathetic letters and merciless messages from the ghostly Red Brigades, which channeled the indignation of the simple people and the poor in spirit, and thus gave some weak probability to the whole story

---

[53] Benigno Zaccagnini (1912-1989), one of the founders of the Christian Democratic Party.
[54] Francesco Cossiga (1928-2010), the Minister of the Interior at the time Moro was kidnapped and, in 1979, the Prime Minister of Italy.
[55] *Note by Jean-François Martos*: Allusion to Moro's defense of the secret services to Parliament during its investigation of General De Lorenzo's failed coup d'état of 1964.

and a reason for it to manifest itself as a collective psychodrama. The general contemplation and passivity continued to hold, which was the most important thing.

If Moro had been killed along with his police escorts on the Via Fani, everyone would have thought it was just another settling of accounts between the capitalist *gangs* and rival centers of decision-making – *which is actually what it was.* In that case, the death of Moro would have been regarded like the death of Enrico Mattei,[56] neither more nor less. Yet no one has noted that, if some powerful group today found that it was necessary or in its own interests to eliminate someone like Mattei or Kennedy, this group would certainly not do it the same way that it had been done in 1962. Instead, they would claim the attack or have it claimed (in a secure way and with the greatest ease) as an assassination by this or that small and secret terrorist group. This is why, in the case of Aldo Moro, one had to stage a long, drawn-out kidnapping: to emphasize the sometimes pitiful, sometimes pathetic, sometimes "firm" character of the government and, when one calculated that the people were convinced of the "revolutionary" origin of the kidnapping and the responsibility of "extremists" for it, then and *only then* did Moro's captors receive the "green light" to get rid of him. And you, Andreotti, who are less naïve than you are flippant, don't tell me that all this is news to you, and do not feign offended virtue, if you please!

The cloud of smoke raised in the country, which concerned the question of knowing if one had to deal [with the kidnappers] or not – a question that still impassions many cretins –, was the thing that had to succeed the best and was, on the contrary, the thing that failed the worst. Here the artificial aspect of the entire machination, put onstage from just behind the scenes, appeared even more clearly than the staging of the kidnapping itself. The people who refused to negotiate, that is to say, the leaders of the Christian Democratic Party and the Italian Communist Party, refused to do so because they knew perfectly well that the staging of the drama foreshadowed the epilogue that was actually offered to us, and because they also knew that, given the situation, they couldn't lose the opportunity to *for once*[57] appear inflexible at

---

[56] Enrico Mattei (1906-1962) was the administrator of Italy's National Fuel Trust. He was killed in a mysterious plane crash that was originally investigated (and found to be an "accident") by Giulio Andreotti, then the Minister of the Interior. The crash was reclassified as a murder in 1997, but no suspects have ever been identified.

[57] Latin in original.

the expense of others. This is why we can admire Zaccagnini and Cossiga, Berlinguer and Pecchioli[58] gargling unrestrainedly with the phrase "dignity of the republican institutions," which had already been so well respected by then-President Leone.[59] The leaders of the parties that refused to negotiate also knew that they could not lose the opportunity to see Moro dead, and thus much less dangerous to them than alive, because a dead friend is much more valuable than a living enemy. Hypothetically, if Moro had been freed, which was impossible, the Stalinists and the Christian Democrats knew quite well that Moro would be three times more dangerous to them than if he were dead: his popularity would be reinforced by his adventure; he'd been discredited in every way by his "friends" when he couldn't defend himself; and thus he'd be an open [and popular] enemy of both his "friends" and his former Stalinist allies. Thus, given the situation, no one has the right to criticize Andreotti and Berlinguer, because they only acted in their own best interests. What one can reproach them for was having done it *so badly,* that is to say, for having raised more doubts and suspicions than applause through their sudden and unforeseen conversion to an inflexibility that obviously did not derive from their respective characters, their past histories, nor their alleged will to safeguard the institutions, which their deeds scorned at every instant, and so this inflexibility had to derive from their undisclosable [and true] interests.

As for Berlinguer in particular, he did not lose the opportunity to once more show himself (as if everyone had not already been convinced) to be the most inept politician of the century. In fact, from the beginning he was as clear as day that the kidnapping of Moro was above all a blow *against the "historic compromise,"* and certainly not dealt by Leftist extremists – who, in any case, would have kidnapped Berlinguer himself to punish him for his "betrayal" – but a group of powerful and interested people who were irrationally hostile to the "compromise" with the so-called Communists. I say *irrationally* because such a policy would certainly not undermine the interests of capitalism. But obviously diligent Berlinguer was not successful in convincing *all* the political sectors, military circles and powerful groups of this, despite the fact that he dedicated five years to this task and to this task alone. And so Aldo Moro, for a long time designated as the artisan of the government of "national unity," paid

---

[58]Enrico Berlinguer (1922-1984) was the National Secretary of the Italian Communist Party from 1972 until his death. Ugo Pecchioli (1925-1996) was the head of the Italian Communist Party's National Commission.

[59] *Note by Jean-François Martos*: Forced to resign due to charges of corruption lodged against him.

the price just as he was bringing that enterprise into port. As Machiavelli said, "from which one draws a general rule, which never or rarely fails: that whoever is the cause of another becoming powerful is ruined."[60] And it isn't by chance that he makes this remark in the chapter entitled *De principatibus mixtis* and that the current governmental majority is also mixed. With the disappearance of Moro, all the other political leaders who had been partisans of the "opening," Democratic Christians and others, were *warned*, because those who decided upon and put into operation the kidnapping of Moro thereby demonstrated that, at any moment, they could do even *worse*. Craxi[61] was the first to understand this, but [eventually] all of the politicians did. And Berlinguer, instead of denouncing this immediately, instead of admitting that the blow struck his politics dead, once again preferred to keep quiet, feigned to believe all the official versions of the facts, played the zealot in the hunt for witches, incited the population to become informers (one doesn't know about what or whom), continued to pad out his own lies, supported Christian Democratic intransigence and hurled invectives against the extremists in the naïve illusion of thereby reassuring the hidden sectors that had kidnapped Moro. But the strategists behind the Via Fani operation mocked Berlinguer's abstract good will against subversives, because they knew that he knew and because they also knew that, when it is a question of real subversion, which harms the economy, Berlinguer could do nothing at all to prevent the actions of the wildcat workers. It isn't enough to *want* to defeat subversion, Berlinguer: you must also demonstrate *that you can do it*. The leaves of abstract [good] will are made of dry leaves that have never been green, imbecile!

And, in fact, as everyone can determine, the Italian Communist Party [ICP] hasn't ceased since then to experience the bitter consequences of its stupid dishonesty. During the kidnapping, the ICP was widely accused by the bourgeois press of definitively being the ones responsible for it because the so-called Communists had encouraged all sorts of illusions about the social revolution among its militants and obtained beautiful results from doing so. Then it lost the elections; then abject Craxi (who during the abduction had already had his eye on the side of those in favor of negotiation, which he knew was impossible, but which permitted him to differentiate himself from the others [in his party]) passed over to the offensive by accusing the Stalinists of everything, but dressed these accusations up under the cover of heated

---

[60] Machiavelli, Chapter III, "Of Mixed Principalities," *The Prince*.
[61] Benedetto Craxi (1934-2000) was the head of the Italian Socialist Party.

ideological quarrels that served as pretexts, which were all the more laughable because they came from a man of his intellectual and cultural stature. But each time the one who lost these quarrels was Berlinguer, and the ICP – because it had not wanted to be fought by its allies in the government – had also forgotten how to fight them. Upon each defeat that it suffered, one witnessed the passably comic scene in which Piccoli[62] and Andreotti caressed Berlinguer's neck, and recommended that he not despair and continue on as before. And yet, despite all these reversals, even today the Stalinists stubbornly continue to feign to believe that Leftist extremists killed Moro. Thus one can say that the interminable series of failures that the ICP has incurred has been *truly merited,* since it is nothing as "the party of struggle" and nonexistent as "the party of government." That which appears to me less comprehensible and more unjustified than all the rest is the fact that the Stalinists lament these failures without any modesty and always portray themselves as victims, but without ever saying *what* they are the victims of, that is to say, their own inaptitude, on the one hand, and the intrigues of their enemies, on the other, and these enemies are much less inept and indecisive than they are, as the Via Fani operation, among others, testifies to and certifies.

The parties in favor of negotiation, on the other hand, survived their defeat, and drew some strength from the weakness of the parties opposed to it. The former were represented by Craxi for reasons of pure convenience and by Lotta continua[63] due to the extreme stupidity that prevented even these militants from perceiving that they are *an integral part* of the spectacle that they want to combat and with which they feed themselves with both hands. Naturally, in this party in favor of negotiation there were many intellectuals, whose perspicacity and depth of thought no longer need demonstration. In any case, these characteristics were supplemented by the crassest ignorance of history, which is even less pardonable on the part of those who have a comment to make about everything and make money from their alleged knowledge. Let me explain: that which above all unites bourgeois reactionaries, the good souls of the progressive bourgeoisie, fashionable intellectuals, the contemplative supporters of armed struggle and the

---

[62] Flaminio Piccoli (1918-2000) was a member of the Christian Democratic Party.
[63] *Lotta continua* ("The Struggle Continues") was a far-Left extra-parliamentary organization founded in 1969 and disbanded in 1976. Its self-titled publication continued until 1982.

militants who complain about it is precisely the fact that, apropos of Moro, they all believe that, for the first time, the State *hasn't lied* where an act of terrorism is concerned, and *therefore* the kidnapping was the work of revolutionaries, with respect to whom the lugubrious Toni Negri[64] has said, "we underestimated their effectiveness (...) We are disposed to make our self-critique" for having "underestimated their effectiveness." Thus, all these people, willingly or unwillingly, are the victims of this umpteenth lie by the State. Both the extra-parliamentarians and the Leftist intellectuals certainly admit that the State always makes use of terrorism *after the fact*,[65] but they cannot conceive that it would also have recourse to killing its "most prestigious" representative. And this is why I spoke of their ignorance of history: none of them know or, in any case, none of them remember the infinite number of examples in which States in crisis, in *social* crisis, have precisely eliminated *their most reputable representatives* with the intention and in the hope of arousing and *channeling* general indignation – generally ephemeral – against "extremists" and malcontents. Of a thousand possible historical examples, I will only cite the Czarist secret services, the formidable Okhrana, which – foreseeing with terror (and with good reason) the revolution of 1905 – killed no one less than Plehve, the Minister of the Interior, on 28 July 1904 and, when this didn't seem sufficient, killed Grand Duke Serge, uncle of the Czar, a very influential man and the head of military conscription in Moscow, on 17 February 1905.

These perfectly successful attacks were organized, executed and claimed by the "Combat Organization" of the Revolutionary Socialists, who had just come under the direction of the famous Azev, a truly ingenious engineer and Okhrana agent, after he replaced the revolutionary Guerchuni, who was opportunely arrested shortly before.[66]

I cite this unique but admirable example of provocation because five

---

[64] Antonio Negri (born 1933) is a Marxist sociologist and philosopher. He founded *Potere Operaio* ("Workers' Power") in 1969 and was a leading member of *Autonomia Operaio* ("Workers' Autonomy").

[65] Latin in original.

[66] *Author's note*: arrested thanks to Azev, Guerchuni heartily recommended to his comrades that Azev himself should be placed at the head of the "Combat Organization" due to the courage and daring he showed while transporting weapons, explosive and publications of the Revolutionary Socialists into Russia from Switzerland, where this party's Central Committee was in exile (specifically, in Geneva).

hundred pages wouldn't be enough to cite all the notorious examples from the 19th century, and because Italy in 1978 had a vague but quite real resemblance to Russia in 1904-1905. In any case, we must note that all powers in difficulty *always resemble each other,* just as their behaviors and manners of proceeding [in such instances] always resemble each other.

The logic currently followed by the strategists of this [terrorist] spectacle is simple, flat and ancient. Provided that we do not recognize their *real* difficulties or the irremediable contradictions with which this old society struggles, the masters of the spectacle of terrorism can flatly present to us the most contradictory things: the terrorism of 1978 is presented as the unavoidable consequence of the proletarian revolts of 1977 and [the bombing of] the Piazza Fontana is presented as the logical end of the "hot" year of 1969. Nothing is more false! The revolts of 1977 were [in fact] the consequence of the "hot" autumn [of 1969] and the kidnapping of Moro was [in fact] the consequence of the provocation of the Piazza Fontana. History advances through dialectical contradictions but, like the scholastic philosophers, the spectacle flatly proclaims *post hoc ergo propter hoc* (after this fact, therefore because of this fact): *the fault* is attributed to the fact. In 1977, the young proletarian generation rose up against its misery? Well, [that means] in 1978 these same enraged young people kidnapped Moro! And it hardly matters that the Red Brigades [RBs] had nothing to do with the revolt of 1977, which they, on the contrary, accused of "spontaneity-ism": the young proletarians of 1977 were subversives; the RBs are made up of young people; [therefore] the RBs are the subversive elements of 1977. Not at all, gentlemen of the government! And you, the general officers of the unofficial services, *since you are always deceived,* you would like it if everyone were just like you! And whoever denounces your provocations is immediately accused of being the provocateur, because reality is always upside-down in the spectacle.

Gentlemen of the government, the truth is that, as in 1977, when your chairs shook under your asses, and the earth shook under your feet, *you* – yes, precisely you – went on the counter-offensive, only this time you killed one of your own, precisely the one whom you and your secret auxiliaries considered to be the most able to arouse popular indignation (no one would have raised an eyebrow if it had been Rumor or even Fanfani[67] who had been kidnapped),

---

[67] Mariano Rumor (1915-1990), a member of the Christian Democratic Party, was the Minister of the Interior in 1963 and between 1972 and 1973, the Prime Minister of Italy between 1968 and 1970, and then again between 1973 and 1974, and the Minister of Foreign Affairs between 1974 and 1976. In

the one who was the most responsible for the current "political framework," which, as you can see, did not please *all* of the capitalist sectors that you and your military organizations are called upon to defend. In his circumstance, one can say that Moro was the Italian equivalent of Allende,[68] and behind the [false] accusation that Moro was serving the interests of the bourgeoisie and capital instead of those of the proletariat, there was in fact (and badly camouflaged) *the opposite accusation,* that is to say, the accusation that Moro wasn't serving capitalist interests in the way that certain capitalists had wanted.

On 16 March [1978], the day of the Via Fani operation, I could not stop myself from immediately thinking two things. First, I thought that the secret services had *finally* been reorganized and had recovered a bit from the affair of 12 December 1969 and the humiliations that followed from it[69] (here again and once more, reality is inverted in the spectacle: one attributed the success of the Via Fani operation to the non-existence of the secret services).[70] Second,

---

1973, Rumor was the target of a bomb that was set by Gianfranco Bertoli, allegedly an anarchist but actually an agent for the *Servizio di Informazioni delle Forze Armate* (SIFAR). Amintore Fanfani (1908-1999), a member of the Christian Democratic Party, was the Prime Minister in 1954, between 1958 and 1959, and then again between 1960 and 1963, as well as the Minister of the Interior between 1954 and 1955, and the Minister of Foreign Affairs between 1958 and 1959, in 1965, and between 1966 and 1968.

[68] Salvador Allende Gossens (1908-1973) was the President of Chile between 1970 and 11 September 1973, when he was deposed and murdered by the Chilean military, with the support of the American Central Intelligence Agency.

[69] Formed in 1966, the *Servizio Informazioni Difesa* ("Defense Information Service") was officially dissolved in 1977.

[70] It is our duty to point out to the reader that these statements are not in accord with what Sanguinetti wrote in a letter to Guy Debord dated 1 June 1978, published in Editions Champ Libre *Correspondance, Vol. II* (1981): "On 16 March, the day Moro was kidnapped, I was in Milan, where I had a meeting with the Doge [Ariberto Mignoli] in the afternoon. In the morning, when the news of the event in Rome echoed on all the streets of Italy, chance would have it that I met Pietro Valpreda, who I immediately asked if this time he could come up with a better alibi than before. Since he said that he didn't have one, and I didn't either, I told him that nothing could be better for me if we were seen *together* on that morning, because no one would bother me if I

I thought of the passage in *Candide* where it is stated that, "in this country, it is good, from time to time, to kill and admiral to give courage to the others."[71]

Sciascia, who is the best known of the Italian readers of Voltaire, certainly isn't the most subtle one, since – forgetting about this passage and all of reality – he lost himself in this or that phrase from one of Moro's letters without discovering that no detail observed under a microscope can indicate or let one catch a glimpse of *the entirety* of the facts. And indeed, even today, Sciascia believes that Craxi and the others really had an interest or the intention of working with "the revolutionaries" [to free Moro] and, with the eloquence worthy of the best defense attorney, he is indignant about the lack of friendship shown for Moro by his "friends," which is an insignificant detail, instead of reserving his indignation *for what is essential,* namely, the facts that virtually the entire world was deceived by this provocation, [new] police-related laws were passed and, despite the hypocritical and despicable appeals from the intellectuals and the pope against "extremism," a hundred innocent people are now locked up in prison for a very long time. Tell me, Sciascia: what importance does it have for history, or even for the truth of the matter, that Aldo Moro had, among others, the misfortune of having such disloyal and dishonest "friends"? Perhaps it is a novelty that the Roman political world is

---

could prove – in any situation that could arise – that I was with a person with a completely *burned* reputation, and thus no one would dare to disturb me a second time. He then invited me back to his place to listen to the first news reports, and it was there that I proposed to him – since he is so well known in the entire world in connection with the provocation of 1969 – that we immediately make a public, printed declaration in a completely sarcastic tone that he cheerfully 'claimed' responsibility for this new provocation, since it clearly came from the same people who placed the bomb in the Piazza Fontana. I even wrote a short text for him, but as you know he isn't the boldest man in Milan, nor the most lucid, and thus he refused it in a categorical manner, with the argument that he'd had his fill of prisons, police and provocations. He offered me a small bottle of Barbera, which, beyond an alibi, was the only thing he offered me (...) Italy's terrorists are not eagles, but its secret services are *nonexistent* (crushed under the weight of their 1969 attack, the arrests in Catanzaro, and the dismantling undertaken by Andreotti himself). (...) The Italian secret services have been *sure of being the only ones* to commit terrorist attacks for such a long time that, when real terrorism takes places, they've been taken completely by surprise."

[71] Voltaire, Chapter 23, *Candide, or the Optimist.*

made up of scoundrels and assassins? Sciascia, have you never read what Cardinal de Retz (a better pamphleteer than you) said three centuries ago? "There are many people in Rome who would be happy to assassinate those who are [lying] on the ground." You, the new Emile Zola, do not accuse the enemies of Dreyfus,[72] but his calumnious friends; not the criminals and the ones responsible, but those (they abound among the journalists for *Corriere della Sera*, for which you write) who have the simple fault of calumnying and dishonoring the victim, *after the fact*.[73] Sciascia, if you regret the fact that Moro had such "friends," why don't you begin by setting a better example, by ceasing to fraternize with the indecent and unspeakable Bernard-Henri Lévy?[74] But I have already said the unsayable about the intellectuals, and it is useless to add any more.

As for the groupuscules with revolutionary pretentions, which have all thrown themselves headlong into theological dissertations about violence and the strategy of "revolutionary" terrorism, I will only recall here that they have long since proved the nature of their comprehension of reality, starting with [the bombing of] the Piazza Fontana, then on every subsequent occasion, such as when they applauded the assassination of Calabresi[75] without stopping to think that this police commissioner had been eliminated *by his own bosses*, for whom he had become cumbersome (he had been involved in the prosecution of Valpreda, the assassination of Pinelli, and something else: several weeks before he was killed in his turn, he had "recognized" Feltrinelli[76] in the unrecognizable cadaver found in Segrate, something for which all the newspapers celebrated "his memory, his shrewdness," etc. without any of them wondering if he managed to do this thanks to his [keen] memory, his

---

[72] Emile Zola (1840-1902) was a French author. On 13 January 1898, the newspaper *L'Aurore* published his open letter, entitled *J'accuse!* ("I Accuse"), which concerned the conviction of Alfred Dreyfus, a French soldier falsely accused of espionage.

[73] Latin in original.

[74] Author of *La Barbarie à visage humain* (1977), an anti-Marxist diatribe.

[75] Luigi Calabresi (1937-1972) was a high-level political police officer in Milan who was tasked with investigating the bombing at the Piazza Fontana. He was murdered on 17 May 1972, allegedly by members of *Lotta continua.*

[76] Giangiacomo Feltrinelli (1926-1972) was the founder of a publishing house and a member of the *Gruppi di Azione Partigiana* ("Partisan Action Group"). He was killed on 14 March 1972, allegedly while setting up explosives underneath an electrical pylon.

shrewdness or, on the contrary, *something quite different*[77]).

These alienated extra-parliamentarians always lose themselves in everything that the Stalinists say about terrorism because they do not know that the ICP can only *lie* and the only thing they can never believe is *the simple truth*: for example, that the RBs are masterminded, that Moro was eliminated by the unofficial services, or that they themselves are fucking idiots, good to throw into prison any time it is useful to do so.

The Stalinists, from the moment that they can be [justly] accused of not knowing what is fascist, or not being able to distinguish what is simply *police-related* from what is fascist, must be accused of lying when they say that the provocation of the Piazza Fontana was "fascist style," and they lied quite maladroitly, because they didn't say "this is fascist," but "this is fascist *style*." The fact that General Miceli,[78] openly fascist today,[79] was already a fascist when he was the head of the SID did not determine his actions back then: the secret services receive their orders from the politicians and do what the politicians tell them to do. Though maladroit, the Stalinists' lies about the bombing of the Piazza Fontana certainly had motivations behind them. Because they wanted to keep quiet about what they knew, and because they, too, were attacked (and quite violently) by the wildcat workers, the Stalinists had to give credence to the ghostly "fascist danger" of 1969, in the face of which they could try to reconstitute "the unity of the working class" under their control. A week after the bombing, metalworkers in the private sector, who were in the forefront of the [proletarian] movement and were its toughest part, were forced to give up their right to strike (starting with the one announced for 19 December) and to accept the contract imposed on them by the unions. Longo[80] and Amendola knew quite well that, if they had immediately told the truth, *the civil war would have begun on 13 December,* and today they know that those who try to be invited to eat at a corner of the State's table can certainly not say out loud that *the plates are dirty,* and so they

---

[77] i.e., he had received orders to lie.

[78] General Vito Miceli (1916-1990) was the head of the *Servizio Informazione Difesa* between 1970 and 1974. He was arrested and imprisoned in October 1974 on charges that he participated in the failed Borghese *coup d'état* of 8 December 1970. He was acquitted in 1978.

[79] He was a member of the Italian Social Movement, founded in 1946 by supporters of Benito Mussolini.

[80] Luigi Longo (1900-1980) was a member of the Italian Communist Party from 1964 to 1972.

say, quietly and secretly, "the plates are dirty, we know, but if you invite us, we will keep quiet about it," which is precisely what has happened.

Since the Stalinists kept quiet in 1969, this so-called "party of clean hands" had *to continue to keep quiet* and lie about all the subsequent provocations and assassinations perpetrated by the secret services of the very State from which, today, they want to receive recognition for observing the *omerta* and, as payment, a few crumbs from the Christian Democrats.

For a long period, the situationists were the only ones in Europe to denounce the Italian State as the creator and exclusive beneficiary of modern, artificial terrorism and its entire spectacle. And, to the revolutionaries of all countries, we identified Italy as the European laboratory for counter-revolution and the privileged field for experimentation with modern police techniques, and we did so starting on 19 December 1969, when we published our manifesto entitled *Is the Reichstag Burning?*[81]

---

[81] *Author's note*: this is the occasion to cite, as an example of revolutionary lucidity, several passages from this manifesto, which one could find posted at the Piazza Fontana and the principal Milanese factories during the period when the repression was the worst. "(...) Faced with the rise of the revolutionary movement, and despite the methodical recuperation undertaken by the unions and the bureaucrats of the old and new Left, power saw itself constrained (...) to play the false card of terrorism (...) The Italian bourgeoisie of 1969 (...) no longer needs the errors of the anarchists of the past to find pretexts for the political realization of its totalitarian reality, but instead seeks to manufacture such pretexts on its own by cornering the anarchists of today in a police machination (...) *The bomb in Milan exploded against the proletariat.* Intended to strike the least radicalized categories and thus ally them with power, and to give the call to arms to the bourgeoisie (...) It isn't at all by chance that there was a massacre among the farmers (at the National Agricultural Bank) and only the fear of one among the bourgeois (the unexploded bomb found at the Commercial Bank). The direct and indirect results of the attacks *were their purpose* (...) But the Italian bourgeoisie is the most miserable in Europe. Incapable of making its own active terrorization of the proletariat succeed, it can only attempt to communicate to the majority of the population its own passive terror, that is to say, its fear of the proletariat. Powerless and maladroit in its attempts to stop the development of the revolutionary movement and, at the same time, [unable] to create a strength that it does not possess, the Italian bourgeoisie risked losing both battles on a single blow. Thus, the most advanced factions of power (internal or unofficial)

The final phrase of this manifesto – "Comrades, do not let yourselves stop here" – is, without exception, the only thing that has been challenged by [subsequent] history. The movement stopped on that precise day and it couldn't be otherwise, because we were the only ones who had full awareness of what the Piazza Fontana operation meant and *we said what it was*,[82] without any other means than a "stolen mimeograph machine," as was indicated in our manifesto. As the people say, "those who have bread have no teeth, and those who have teeth have no bread." All those courageous extra-parliamentarians who had newspapers and other rags had no teeth, and they published nothing pertinent about the massacre, occupied as they were, and still are, with the search for the "correct strategy" to impose on the proletariat, which is only good for being directed and being directed *by them*!

Because of their incurable inferiority complex concerning the ICP's ability to lie, which is indeed superior to theirs, these extra-parliamentarians immediately accepted the version of the facts accredited by the ICP, according to which the bombs were "fascist style" and *therefore* could not have been the work of the secret services of this "democratic" State that is so "democratic"

---

have made a mistake. Excessive [social] weakness has brought the Italian bourgeoisie onto the terrain of police excess: it understands that its only possibility of getting out of its endless agony passes through the risk of the immediate end of that agony. Thus, right at the start, power has had to burn the last political card it has to play before [the outbreak of] civil war or a *coup d'état* of which it is incapable [of winning or defeating] – the two-faced card of a false 'anarchist peril' (for the Right) and a false 'fascist peril' (for the Left) – with the goal of masking and making possible its [counter-]offensive against the real danger: the proletariat. Moreover, the act with which the bourgeoisie has tried to avert civil war is, in reality, its first act of civil war (...) Thus, it is no longer a question of the proletariat avoiding or beginning it, but winning it (...) The proletariat now begins to understand that it isn't by partial violence that this civil war can be won, but by the total self-management of revolutionary violence and the general arming of the workers organized into Workers' Councils. It now knows that, through revolution, it must definitively reject the ideology of violence as well as the violence of ideology (...) Comrades: do not let yourselves stop here (...) Long live the absolute power of the Workers' Councils!"

[82] *Author's note*: the only exception to the general rout was "Bombs, Blood, Capital," a tract by Ludd, published in January 1970, that openly accused the secret services of the massacre.

that it never worries about is said by these extra-parliamentarians, although they are the only ones considered to be "dangerous" to the spectacle, for which they are badly compensated but indispensible walk-on actors. Their false explication of the facts perfectly matched the true ideology of their groupuscules, then infatuated with Mao, Stalin and Lenin, and now by Guattari, Toni Negri and Scalzone,[83] or by their miserable "private lives" and ridiculous whorehouses. Since these alleged "extremists" *do not want* to tell the truth, and do not know how to openly accuse the State of being *the* terrorist, they also do not know how to combat it with any tangible results. Because saying that the bombing was "fascist" was as mendacious as saying that it was "anarchist," and all the lies – though apparently in contradiction with each other – are *always united* in the sabotage of the truth. And only the truth is revolutionary; only the truth is able to *harm* power; only the truth can infuriate the Stalinists and the bourgeois. And the proletariat, always deceived and betrayed by everyone, has learned to seek the truth on its own and is impervious to lies, no matter how "extremist" their authors claim to be. In the same way, and due to the same *guilty* ineptitude, all the extra-parliamentarians of 1978 merrily fell into the trap set by the kidnapping of Moro, "the work of comrades who were mistaken." You great oafs, don't you realize that, once again, you were the only "comrades who were mistaken"? Brave extra-parliamentarians, Dante already wrote your epitaph.

> But you take the bait, so that the hook
> Of the old adversary draws you to him;
> And so check and recall do very little.[84]

Victims of their own false consciousness, which always expresses itself in ideology, the extra-parliamentarians could not for long elude the questions posed by spectacular terrorism and, from 1970 on, they began to consider the question of terrorism as such, in the empyrean of ideology, in a completely metaphysical way, completely abstracted from the reality of the thing. And when the truth about the massacre at the Piazza Fontana finally saw the light of day, after all the lies about it collapsed one after the other, neither the good

---

[83] Pierre-Félix Guattari (1930-1992) was a French militant, psychotherapist and philosopher, perhaps best known for his collaborations with Gilles Deleuze. Oreste Scalzone (born 1947) is a Marxist intellectual and one of the founders of *Potere Operaio* ("Workers' Power").

[84] Dante, *Purgatorio*, XIV, 145-147.

souls of the intellectual-progressive bourgeoisie nor the scarecrows for sparrows at Lotta continua and their consorts were able to pose the question in its real, that is to say, *scandalous* terms: *the democratic Republic [of Italy] did not hesitate to perpetrate a massacre* when it appeared useful for it to do so, because, when all the laws of the State are in danger, "there is only a single and inviolable law for the State: that of its survival" (Marx).[85] And this is *exactly* the famous "sense of the State"[86] that one made Moro assume and with which the philistines now decorate his corpse. In ten years, no one has wanted to trigger a "Dreyfus affair"[87] concerning the behavior of our secret services, the leaders of which enter and exit prison on the sly, to the general indifference of all the privileged holders of the "sense of the State," that sublime sixth sense with which our politicians are endowed, unlike common mortals, who are mutilated by it, such as those who were at the Agricultural Bank and not killed [on the day of the bombing], but by something else. Perhaps there is someone who is convinced that this mysterious "sense of the State" is something other than what I've said it is. "Moro had the sense of the State" and "Berlinguer has the sense of the State": if these expressions do not mean what I've said they mean, then they are empty, which means that one could say that a young woman has "the sense of the cunt," I have the sense of my balls, and Tina Anselmi[88] doesn't have [much] sense even if she *causes* a sensation.

Since the extra-parliamentarians at first did not believe they knew, then knew without believing, and finally believed *without concluding* that *it was indeed the State that launched the terrorist attack in Milan,* the entire country

---

[85] Karl Marx, "The Trial of the Rhenish District Committee of Democrats," speech delivered on 8 February 1849 and printed in *Neue Rheinische Zeitung* #231-232, 1849.

[86] Cf. Hegel, *The Philosophy of Right.* Note that, in 1958, Giulio Andreotti published a book entitled *Il senso dello stato.*

[87] i.e., trigger a national scandal such as occurred after Emile Zola published his *J'accuse!* letter. Cf. Guy Debord's letter to Sanguinetti dated 21 April 1978, published in Editions Champ Libre *Correspondance, Vol. II* (Paris, 1981): "There wasn't a public 'Dreyfus affair' [over the bombing at the Piazza Fontana], not because the scandal was less great, but because no one ever demanded a true conclusion. Thus Italy, which has experienced a 'creeping May [1968],' has worsened its sickness by a 'suppressed Dreyfus affair.'"

[88] A Christian Democrat (born in 1927) and the first female member of an Italian cabinet, first as Minister of Labor, then as Minister of Health.

has entered into a period of apparent madness and mad appearances. The entire question of terrorism has become the subject of academic diatribes and enthusiastic invectives that have led some (the bourgeois and the Stalinists) to hypocritically condemn terrorism "whatever its color" – as if they weren't precisely the ones who have encouraged and covered it up, each time, by giving it *the color that best suited the moment* – and have led the others (those who believe themselves to be "extremists") to fondle the idea that "one responds to State terrorism with proletarian terrorism." And this comes at the right time for our secret services. The first small, clandestine terrorist groups (the RBs and the Armed Proletarian Nuclei [APN]) had just been formed when the police, the Carabinieri and the detached units started competing to see which one could be the first to infiltrate these small para-military groups with the goal of preventing their attacks or masterminding them according to the necessities and *desires* of the moment and the powerful.

Thus everyone could see how the APN were radically destroyed, either [indirectly] by arresting their members and exhibiting them in a disgusting way at this or that trial, or directly by turning them into objects for target practice, a meticulously arranged spectacle in which the "forces of order" were exhibited for the pleasure of the most repugnant bourgeois.[89]

Things panned out differently with the Red Brigades. We know the names of only two of the agents who infiltrated this group, that is to say, Pisetta and the Christian Brother, Girotto,[90] who – despite being quite clumsy as *agents provocateurs* – were able to trap Curcio[91] and the other members of what can justly be called the "historical group" (the militants who had no experience with clandestinity and were hardly "ferocious" as terrorists). Despite this, the RBs were not dismantled after being decapitated [in September 1974], not because of the prudence of the other militants, who

---

[89] *Author's note*: this bloody spectacle was offered sparingly, but in a repeated fashion: when the police waited for Abatangelo in front of the Bank of Florence and killed two of his comrades; when Mantini's sister was killed in cold blood in her secret hideout in Rome; and dozens of other cases. Should one believe that it was by chance, and not due to infiltration, that "Italy's Finest" obtained such *successes*?

[90] Marco Pisetta (1945-1990), who knew Renate Curcio back in 1968, infiltrated the Red Brigades in 1972. Silvano Girotto (born 1939) was also known *Frate Mitra* ("Brother Machinegun"). Working with General Dalla Chiesa of the Italian Carabinieri, he infiltrated the Red Brigades in 1974.

[91] Renato Curcio (born 1941) co-founded the Red Brigades in 1970.

were no less naïve than their original leaders (who themselves fell into the very first trap set for them), but *because of the decisions made by their new leaders.* And why would the State, already in difficulty *for other reasons,* lose this opportunity to make use of a terrorist organization that had an autonomous appearance, although infiltrated and tranquilly directed from afar? I do not at all believe that General Dalla Chiesa[92] is the "warrior genius" of which Carl von Clausewitz once spoke, but he certainly read Clausewitz with more attention and profit than Curcio did and [in any case] had greater means to put at the disposal of his talents. General Dalla Chiesa – along with his colleagues at the SISDE, the SISMI and the CESIS[93] – had a good laugh at all the proclamations of the ideologues of armed struggle who intended "to bring the attack to the heart of the State," because Chiesa knows that the State doesn't have a heart, not even a metaphorical one, and because, like Andreotti and Berlinguer, he knows that *the only attack capable of killing the State today is the violent denunciation of its terrorist practices,* which is precisely what I am making here.

Although he is better informed about tactics than strategy, and although he confuses strategy with stratagem, thus substituting cunning for the art of war, General Dalla Chiesa nevertheless knows perfectly well that terrorism is *the substitute for war* in an era in which large-scale world wars are impossible or, in any case, it is no longer permitted to have one proletariat massacre another in an exhausting and bloody battle. Our General and the upper-level strategists of the political police also know that spectacular terrorism *is always anti-proletarian* and that it is *the pursuit of policy by other means* (the pursuit of the anti-proletarian policy of all the States). And the fact that the State *needs* modern, artificial terrorism is proved by the fact that it was precisely here, in Italy, that the State *invented* this form of terrorism ten years

---

[92] General Alberto Dalla Chiesa (1920-1982) was a high-ranking officer in the carabinieri. In September 1974, his "anti-terrorist" unit captured Renato Curcio. On 3 September 1982, he was murdered, allegedly by the Mafia.

[93] After the dissolution of the *Servizio Informazione Difesa* in 1977, the *Servizio per le Informazioni e la Sicurezza Democratica* ("Intelligence and Democratic Security Service") took charge of domestic intelligence, the *Servizio per le Informazioni e la Sicurezza Militare* ("Military Intelligence and Security Service") took charge of military intelligence, and the *Comitato Esecutivo per i Servizi di Informazione e Sicurezza* ("Executive Committee for Intelligence and Security Services") took charge of coordinating of the activities of the SISDE and the SISMI.

ago. And we know that the Italian bourgeoisie has long used invention to replace what it lacks in power. It was the Italian bourgeoisie *that invented fascism,* which was so successful in Germany, Spain, Portugal and everywhere else it was necessary to crush proletarian revolution. And the spectacle of terrorism has already been immediately successful for the German government, which does not envy our situation, but envies our imagination, that is to say, the imagination of our secret services, which permits our government to navigate through shit without drowning in it, just as in the 1920s it envied us for Mussolini.

That this [Italian] State has need of terrorism is, on the other hand, something that each one of its representatives is now completed convinced of, through experience if not due to reasoning, and has been so ever since the immediately and miraculously fortunate outcome of the Piazza Fontana operation. The proof is that, *if there has not been* a "Dreyfus affair" where the Piazza Fontana is concerned, this is certainly not because the event was less scandalous, but because all the political parties have, for different reasons, understood that, if this bombing saved the State (which each of them defend in their own way), *the truth* about it was capable, by itself, of definitively *destroying it.* And if there has been no "Dreyfus affair," this is also because, among our enslaved intelligentsia, no equivalent of Emile Zola has ever demanded or wanted to demand *a truthful conclusion* about the bombing. Giorgio Bocca's book on terrorism[94] discreetly begins with 1970 [not 1969] and, as for the other Brahmins of culture, such as Pasolini and Sciascia, they have – in the blinding light of the Reichstag fire – preferred *to chase fireflies,* without even finding any, obviously, since they always discuss the responsibility of pollution for their disappearance and raise pleasantly "polemical" lamentations about it, but without ever denouncing *terrorist pollution,* of which they are both the accomplices and the victims.

I would like it if the unofficial services and the generals – who will read *Remedy for Everything,* or at least the chapter that concerns them, attentively – pay immediate attention to two things that I say to them about the fragility of their strategy. Dalla Chiesa, note, above all, what Clausewitz teaches you in the chapter that he dedicates to the stratagem.[95]

But however much we feel a desire to see the actors in war outdo

---

[94] *Il terrorismo Italiano, 1970-1978* (1978).
[95] Carl von Clausewitz, "Stratagem," Book 3, *On War,* as translated from the German by Colonel J. J. Graham (1832).

each other in hidden activity, readiness and stratagem, still we must admit that these qualities show themselves but little in history (...) The explanation of this is obvious (...) In fact, it is dangerous to detach large forces for any length of time merely for a trick, because there is always the risk of its being done in vain, and then these forces are wanted at the decisive point. The chief actor in war is always thoroughly sensible of this sober truth, and therefore he has no desire to play at tricks of agility (...) In a word, the pieces on the strategic chessboard want that mobility which is the chief element of stratagem and subtlety (...) [Craftiness] does no harm if it does not exist at the expense of the necessary qualities of the heart, which is only too often the case.

The second thing to consider with respect to a strategy that is founded on provocation is as old as the world. It is noted by Seneca – and if I quote him, it is because, as Nero's advisor, he knew about State terrorism and provocations – that it is "easier to not go along this road than, once one has begun, to stop."[96] Like a[n addictive] drug, artificial terrorism needs and requires the administration of ever-larger and *more frequent* doses, *because any future seems evil and already is,* as Dante would say.[97] Redo your calculations, politicians and generals, and you will see that *they are wrong.*

If the State needs terrorism, as I have demonstrated, it also needs to avoid getting caught red-handed every time it uses it, so that its ministers can put up a better front than Rumor and Tanassi did at Catanzaro[98] (their only equals in this are Malizia, Maletti and Miceli).[99] And for the State what better occasion than that offered by a group like the Red Brigades, decapitated and available, with its former leaders in prison and ignorant of everything? Nevertheless, I must say that, even if its former leaders were freed – given that

---

[96] Sanguinetti also mentions this remark in his letter to Debord dated 1 June 1978 and published in Editions Champ Libre *Correspondance, Vol. II* (1981).

[97] Dante, *Purgatorio,* XX, 85.

[98] In 1974, Catanzaro was the location of the trial of the alleged perpetrators of the bombing of the Piazza Fontana.

[99] General Saverio Malizia (born 1914) was the legal counsel to the Ministry of Defense and the Deputy Prosecutor of the High Military Court. He was convicted of perjury in 1979. General Gianadelio Maletti (born 1921) was the head of counter-intelligence for the *Servizio Informazione Difesa* between 1971 and 1975. He was convicted of falsifying public documents in 1979.

two infiltrators were enough to bring them down – a single one who was less crude than "Brother Machinegun" or Pisetta would have been enough to make them go where one wanted to make them go without arousing any suspicion. I know quite well that the currently known infiltrators, as well as the majority of the *agents provocateurs* at work today, did not invent the butter knife, but our clandestine militants aren't any more subtle than they are, as we have seen. And even if they were all Lenins, as they imagine themselves to be, one would still have to remark that the Bolsheviks were deeply infiltrated several times. Roman Malinovski, worker and Okhrana agent, made a part of the Bolshevik Central Committee, enjoyed the blindest confidence on the part of Lenin, and sent to Siberia hundreds of militants and leaders. To a suspicion expressed by Bukharin, Lenin (according to his wife, Nadiejda Krupskaia) responded that it was "unworthy of a conscious militant; if you persist, it will be you who will be denounced as a traitor." But the case of Malinovski is not an isolated one. Opening the secret archives of the Okhrana in 1917, Lenin was (not without good reason) stupefied to discover that, of fifty-five officially active and regularly paid professional provocateurs, seventeen "worked" among the Revolutionary Socialists, and a good twenty of them shared the job of surveilling the Bolsheviks and Mensheviks, and certainly not among the rank-and-file militants! And Lenin had the bitter surprise of discovering that the provocateurs were always those "comrades" to which he – the man who was so prudent and so expert in matters of clandestinity – accorded the greatest esteem and the greatest confidence because of their service and the boldness they showed on several occasions.

Today, the practices of the Okhrana, which were very sophisticated and refined for the times, are no more than primitive. The modern unofficial services of the State, of *any* State, dispose of a number of means and people of all classes and all social appearances, well trained in the use of weapons and ideas, and often much more *capable* than the naïve militants, who pay the price for it. The organizational form of the political party, which is always hierarchical, is in fact the one that is best suited for infiltration and manipulation, which is exactly the opposite of what the bourgeois press says. All the rank-and-file nuclei, which are made up of clandestine militants, are kept separate from each other and in ignorance of everything, without any possibility for dialogue and debate, and everything functions perfectly due to the blindest [obedience to] discipline and the most expedient orders from an inaccessible summit, which is generally nested in this or that ministry or powerful group. And, if some provocateur ever arouses suspicion, he is always providentially arrested and made a star by the press, which removes him from danger and washes him of the suspicion. Thanks to an unbelievable and "heroic" escape, he can then be put back into action. And often these provocateurs do not come out unscathed.

Thus, here is *one more reason* why I would warn any subversive of good faith about organizing hierarchically and clandestinely in a "party": in certain conditions, clandestinity can be necessary, while all hierarchies always and only benefit the world we seek to bring down. In revolutionary groups that do without militants and leaders, and that are founded *on the qualitative,* infiltration is practically impossible or immediately discovered. "The only limit to participation in the total democracy of the revolutionary organization is the recognition and effective appropriation by all of its members of the coherence of the organization's critique, a coherence that must prove itself in the critical theory properly speaking and in the relationship between this theory and practical activity" (Debord).[100]

In several of the Red Brigades' "hideouts," and this isn't news, abundant amounts of ultra-confidential materials have been recovered, and they contained the locations of police officers, police stations and even government ministries, which, strangely, have never been assaulted and sacked by the RBs. Confronted with such eloquent facts, [sources of] spectacular information have always pretended to explain them by once more emphasizing the ultra-efficient organization of the terrible RBs and, to strengthen this brilliant bit of advertising, they have added to it the "fact" that these clandestine militants – who are hunted, but so widespread – have infiltrated themselves everywhere, even into the police stations and ministries. Confronted with this explication of such a gloomy and maladroitly camouflaged reality, I can only laugh. Once more, the intelligence of fifty million Italians, who are not Germans ready to drink from the poisoned baby's bottle of the television set, is being abused by the *Corriere* and *L'Unita,*[101] and those who attribute such stupidity to ordinary people in fact *only reveal theirs,* which, to surpass so many limits, certainly cannot be so ordinary. Once more, power speaks in counter-truth: it is not the RBs who have infiltrated into the police stations and ministries, but agents of the State, employed by the police stations and ministries, who have been infiltrated into the RBs on purpose, and certainly not only at their summit.

And if, during ten years, the merciless and great struggle against the

---

[100] Guy Debord, "Minimum Definition of Revolutionary Organizations," July 1966, printed in *Internationale Situationniste* #11, October 1967.
[101] *L'Unita,* founded by Antonio Gramsci in 1924, was the official newspaper of the Italian Communist Party.

"monster" of terrorism – a struggle that has been so glorified in words – has only resulted in the hypertrophy of this "monster"; if the trial [of the suspects in the bombing] of the Piazza Fontana has never truly *begun,* this derives from the fact – which is comic or repugnant, I don't know – that those who have always been tasked with this merciless struggle are *the same secret services that have always directed and animated terrorism,* and certainly not because of alleged "deviations" or "corruption," but *simply because they have executed in military fashion the orders that they have received.* And all the militants who have been exhibited in the public cages of the courts, as if they were ferocious beasts, [but who are really] naïve children whom one would like to see grow old in Italian prisons, are always and most assuredly *the least implicated,* even if they have been designated "the leaders" and "the strategists" (nothing is easier than making a naïve fanatic believe that he or she has taken part in this or that operation simply because he or she distributed the tract that claimed responsibility for it).

And our general officers amuse themselves by counting the medals and attestations of high merit that they collect, either by nourishing terrorism or by "discovering" the "guilty parties" at opportune moments.

In this phenomenon, which might arouse the virtuous indignation of hypocrites, there is really nothing new, and it has been repeated for centuries in the eras of corruption and decadence of all the States. For example, Sallust, who is the historian of the corruption and crisis of the Republic of Rome, reports that the censor Lucius Marcius Philippus denounced Lepidus, a general felon, before the Senate in these noble terms.[102]

> I could wish, beyond all things, Conscript fathers (...) that mischievous plots should prove the ruin of their contrivers. But, on the contrary, everything is disordered by factious disturbances, disturbances excited by those whom it would better become to suppress them (...) While you, whispering and shrinking back, influenced by words and the predictions of augurs, desire peace rather than maintain it, being insensible that, by the weakness of your resolutions, you lessen at once your own dignity and his fears (...) What would he have received for good deeds, when you have bestowed such rewards on his villainies? (...) [He takes advantage of your inactivity] which I do not know whether I should not rather call fear or pusillanimity or

---

[102] Sallust, *Histories,* translated from the Latin by H. G. Bohn (1852).

infatuation (...) [Lepidus] thou are a traitor to us (...)[103] You claim to reestablish through such a war the concord *that is rendered vain by the very means by which it is obtained.* What impudence!

The truth is exactly this: the social peace that terrorism can procure is "rendered vain by the very means by which it is obtained," with the difference that, today, *the impudent ones* are the representatives of the Republic and all the orators who fulminate against terrorism in all of their speeches, always feigning not to know what the entire country has been saying since the famous year of 1969. Listen a little to what is said by a modern Lepidus, the honest Leo Valiani,[104] who in the pages of the *Corriere* during July 1978 was not ashamed to regret the "too mild sentences" given to some underling.

> [They] encourage the subversive to persevere, to dare to do even more. We do not ask the judges to condemn someone without being convinced of his [or her] guilt. But when the Republic is grappling with clandestine organizations such as those who sowed death at the Piazza Fontana, as it is at this moment (...) any indulgence concerning those who are active in such subversive organizations is suicidal.

And, in the name of God, what indulgence could surpass that of Valiani, an expert in Stalinist and bourgeois terrorism, fellow traveler of these two forms of terrorism and accomplice of all the lies about them, who still feigns to be ignorant (and he is not the only one in Italy to do so) of the fact that "the clandestine organization that sowed death at the Piazza Fontana" was none other than Admiral Henke's organization, the famous SID, which for the sake of decency – that is to say, *indecency* – today has a different name? And does one want to continue to listen to Valiani's chatter for the next ten years, only this time with respect to the execution of Moro? What parliamentarian, what honorable bastard, among all those who reproach each other for their "indulgences," talking nonsense about "safeguarding the Republic," has dared to expose himself by accusing and *naming* the assassins of 1969?

The fact is that the task of safeguarding this criminal Republic *depends solely on these parliamentarians' ability for cover for those assassins and the*

---

[103] The next two lines do not appear in the translation of this text by H. G. Bohn (1852), nor in the one made John C. Rolke (1921).
[104] Leo Valiani (1909-1999) was a journalist for *L'Espresso.*

killers of Moro, as well as the murderers of Calabresi, Occorsio, Coco, Feltrinelli, Pinelli, et. al. And all the government ministers and honorable parliamentarians know this very well: they continue to keep quiet so they can receive new remunerations that will complete their already substantial share.

Ever since the great fear of 1969, our regime has accorded an immense trust in its senior political police officers and their abilities to always find *technical and spectacular solutions* to all of the historical and social questions that face it. Thus, our regime is in the process of committing the same error made by the Czarist regime, which dedicated all of its attention and care to building the best and most powerful secret police in the world, which is what the Okhrana was at the time. This permitted the Czar to continue to survive day by day, without anything changing for another decade, but his [eventual] fall was only more violent and definitive. As a bourgeois thinker, Benjamin Constant,[105] has said:

> Only an excess of despotism can prolong a situation that tends to dissolve and can maintain under class domination all those who conspire to separate themselves from it (...) Even more harmful than evil, this remedy has no durable effectiveness. The natural order of things takes revenge against outrages that one wants to subject it to, and the more the compression has been violent, the more the reaction will be terrible.

And in Italy, [the effects of] ten years of political-police policy are beginning to make themselves felt, and that includes their harmful and uncontrollable effects. The State still exists, with more authority and a worse reputation than ever, but its real adversaries have grown in number, their awareness has [also] grown and, with it, the effectiveness and violence of their attacks. In the eras in which the police make policy, a complete collapse is always what follows.

Today, sinister Craxi seeks easy applause by feigning to discover that, in Russia, [mere] crimes of opinion are considered to be crimes against the State. Scandalous novelty! Poor Craxi, have you not seen that here in Italy *crimes against the State* are *considered to be crimes of opinion*? Is this not a fact more worthy of your virtuous indignation? Ridiculous man! Who do you want to convince that your soul is immaculate? You, who strut about with your worthy

---

[105] Benjamin Constant (1767-1830). Quote taken from Chapter XIII, *L'Esprit de Conquête et de l'Usurpation.*

colleague Mitterrand: do you believe that we have forgotten that Mitterrand is *a gangster* who, several years ago, paid other, more obscure gangsters to fake an attack against him?[106] Craxi, no one believe you when you proclaim *I am without fault before the throne!*[107] And you, leaders of the [Italian Socialist] Party: you are just like Mitterrand. When it is a rival, and not one of you, who command the attacks, you keep your mouths shut, and you speak of the firmness of the State when it is confronted with your own provocations!

Here is another proof, in addition to all the rest, that in Italy crimes of State are considered to be simple crimes of opinion. In 1975, when, under the pseudonym of Censor,[108] I published *historical* (not legal) proof that it was the SID that perpetrated the massacre at the Piazza Fontana, all the newspapers and journalists widely reported my conclusions, but they were more scandalized by the fact that an anonymous person, apparently close to power, dared to openly accuse the SID than the completely believed fact that the State had organized and executed a terrible massacre so as to emerge unharmed from a very serious social crisis. The journalist Massimo Riva admirably expressed the thinking of all his colleagues when he wondered, in the *Corriere,* what mysterious maneuver among the powerful the Censor affair announced. "What is behind it? The fear of telling the truth? A warning to the regime's big shots?" It wasn't my scandalous statements or conclusions, but my anonymity that caused the scandal. To say it better: the fuss surrounding the identity of Censor only served to mask the scandal of what I denounced. Everyone preferred to advance maladroit conjectures about my identity and thus avoid speaking of what I said. "A warning to the regime's big shots?" According to Riva and the others, this was the crux of the question, and what was scandalous was only *the end of the omerta* among the powerful, and not the crimes they had committed.

But, as always, the best was Alberto Ronchey, who will only astound us when he no longer manages to be astounding. With respect to my proofs, he

---

[106] On 16 October 1959, François Maurice Adrien Marie Mitterrand (1916-1996) – the future President of France and leader of the Socialist Party, then a senator – was allegedly the target of an assassination plot that was in fact organized in cahoots with a right-wing deputy named R. Pesquet.

[107] *Book of Revelations,* 14:5. Latin in original.

[108] *Truthful Report on the Last Chances to Save Capitalism in Italy.* Note: Sanguinetti didn't merely write this document under a pseudonym; he also invented *the persona* of its hypothetical author. In what follows, he seems to forget this second part of his creation.

said, "Whatever the responsibilities and intrigues of the SIFAR-SID[109] or other detached units," *despite them*, "where the bombs [and] events are concerned (…) if one truly believes in a 'State terrorism,' we would be confronted with a criminal system of government and no one should have anything to do with it, neither the Communists, the Socialists nor the others."[110] What is truly unbelievable is not State terrorism, but Ronchey's manner of reasoning. Since he himself, the Communists and the Socialists in fact have "something to do" with such a government, *therefore* (according to Ronchey) we have a sufficient guarantee that State terrorism is not believable and indeed *does not exist,* "whatever the responsibilities and intrigues" of the SIFAR-SID. To reason as Ronchey does: God is believable, *therefore he exists.*[111] [In the contemporary discourses] on the subjects of terrorism and the State, one has the impression of being returned to the discussions about the existence of God and the Devil. Are they real? Do they exist? And, if they do, are they truly believable? Quite wisely, the poet says, *Certainly it was true, but not at all believable to those who weren't masters of their reason.*

I have not managed to understand where the Roncheys of this world hope to arrive with their theological logic. I have never said that the secret services have been behind every attack, given that, today, even a *Molotov cocktail*[112] or an act of sabotage against production is considered to be an "attack," but I did say – and I have been saying it for almost ten years – that all the spectacular acts of terrorism have been masterminded or directly perpetrated by our secret services. And the reader should note that I didn't say "by *the* secret services," which could refer to those of a faraway or exotic country, but *ours*, yes, the secret services of Italy, whose touch and stench, cleverness and clumsiness, tactical ingenuity and strategic stupidity I always recognize.[113]

---

[109] Originally formed in 1949, the *Servizio di Informazioni delle Forze Armate* (SIFAR) became the SID in 1965.

[110] *Author's note*: A. Ronchey, *Accadde in Italia, 1968-1977.*

[111] Sanguinetti had previously made this point in *Proofs of the Non-Existence of Censor, By His Creator* (January, 1976), the document in which he claimed responsibility for "Operation Censor."

[112] English in original.

[113] We must point out that, to the precise extent that Italy was a member of the North Atlantic Treaty Organization, which has always been dominated by the United States and its strategic interests, the reader is justified in believing that the "masterminds" of Italian spectacular terrorism (the SID et. al) were

For example, observe how the SID came to perpetrate the Piazza Fontana operation: by successive attempts and approximations. It decided to perpetrate a massacre of the population and it prepared for it with two general rehearsals: the bombs at Fiore and the bank at the train station in Milan on 25 April 1969, and the bombs on the trains in August of the same year. The secret services thus prepared public opinion and prepared themselves technically with these *backgrounds*.[114]

The kidnapping of Moro was also rehearsed in advance, because our unofficial services, even if they changes their targets, always have the same manner of proceeding, which is something for which Machiavelli would never pardon them. In April 1977, the kidnapping of De Martino[115] was one such rehearsal, and it took place without the spilling of blood, because our secret services never want blood to be spilled during one of their rehearsals. On both 25 April 1969 and later in August, no one was killed. Nevertheless, such rehearsals have *always indicated the target* to be struck: in 1969, it was the population and in 1977-1978, a politician. The *very day* of the kidnapping of De Martino, which was claimed by a hundred ghostly groups, I denounced it as a general rehearsal by the secret services in a poster that was printed and distributed in Rome.[116] The second rehearsal, which indicated the target that had been selected – that is to say, a politician – was the bomb at the office of Cossiga, then the Minister of the Interior, which assured this act of a lot of publicity.[117] Then came the attack against Moro, and blood was spilled, because it was not a general rehearsal.

Under the pressure of the menacing revolts of the beginning of 1977, the secret services, which had always been on their guard and never inactive

---

"masterminded" in their turn by NATO and the USA. It is certainly the case that the American secret services, i.e., the CIA, is also known for its "cleverness and clumsiness, tactical ingenuity and strategic stupidity."

[114] English in original.

[115] Francesco De Martino (1907-2002) was a prominent member of the Italian Socialist Party. On 5 April 1977, his son, Guido, was kidnapped, allegedly by the Mafia. In exchange for one billion lira in ransom money, he was released on 15 May 1977.

[116] *Author's note*: "Notice to the Proletariat About the Events of the Last Few Hours," Rome, 7 April 1977.

[117] Francesco Cossiga (1928-2010), a member of the Christian Democratic Party, was the Minister of the Interior between 1976 and 1978. On 7 April 1977, a bomb exploded outside of his personal residence.

for ten years, began to move decisively in a quite precise direction, and the two aforementioned provocations (which weren't the only ones made by those services) were, nevertheless, the ones that clearly indicated the target chosen and the events to follow.

Thus we can advisedly say that the kidnapping of Moro was *the least unforeseeable thing in the world,* since it was *the least unexpected thing* there where one can do what one wants, that is to say, at the summit of power. First of all, one feared that De Martino, a friend of the Stalinists, would be elected President of the Republic. By making him pay several hundred million lira to regain his son, one destroyed the reputation of this "Socialist." Moro was then publicly designated Leone's successor and, though he was less valuable as a target for ransom-demanding kidnappers than De Martino or Leone, he was the one responsible for the agreement with the Stalinists and, as President, he would have been even more so. Two plus two equals four, even in politics. And so, on *16 March 1978, the President had to die,* to parody the title of a book by Andreotti.[118] Six months after the Via Fani operation, while the anti-Stalinist policies of Craxi first showed themselves, Amintore Fanfani – who is called *The Ghost* in Tuscany – launched his first rigorous attack against the government, the secretariat of the Christian Democratic Party, the "emergency cabinet," and the "agreement" made by Moro, by denouncing "the abuse of unanimity" and the inefficiency of the "equivocal" government of "national unity," and by announcing the surpassing of "a political season" – all of which was applauded by the Craxians and "feared" by the Stalinists. Although Fanfani is, after Berlinguer, the Italian politician who has collected the greatest number of failures, he *is not* a cretin (more intelligent than clever and less shrewd than ingenious), *the Ghost* only drew *political conclusions* from the Moro affair, since it is true that terrorism is the pursuit of policy by other means.

As long as power exists separately from individuals, it will surely not be individuals who are in short supply. No functionary of power or capital is irreplaceable or indispensible for the maintenance of domination: neither Kennedy, Mattei, Moro, nor any of those who are still alive and active. In periods of trouble, the thing that is indispensible to a power that does not want to be replaced is precisely *the elimination* of certain men, either because they are too compromised or exposed, like Rumor, or because they desire a "replacement" (even a minimal one) that arouses the fears or suspicions of

---

[118] *Ore 13: Il Ministro deve morire* ("1 pm: The Minister Must Die"), published in 1974.

certain sectors of power, and one knows that the most reactionary sectors are always *the best armed*. Moro's "overtures" [to the Stalinists] were thus perceived as opposed to certain interests and as a concession to "change" – and this despite the fact that, historically, it is precisely change that such overtures *try to prevent*, but without too much conviction or sufficient guarantees – that is to say, in a manner different from the one desired by a faction of power and certain military leaders.

In history, all powers have always behaved like all the other powers have behaved and, to the extent that the current police-politics of provocation follows its course (and I have already demonstrated that it cannot fail to do so), its powerful, semi-lucid, semi-unaware and *completely* fear-dominated strategists find themselves with the *necessity* of getting rid (mafia-style) of certain men of whom they made use just the day before. There is nothing new in this, and it is a supplementary confirmation of the old [and previously mentioned] precept, according to which "he who causes another to become powerful ruins himself." Neither Moro nor any of his colleagues prevented the political police from becoming powerful over the course of the last ten years; none protested against or combated a phenomenon that they all, on the contrary, nourished. Moro was the first victim of some importance mowed down by this politics, but he wasn't the only one. The strategists of terror had already gotten rid of other, less important, but no less utilized people. We can cite several still-fresh examples: the liquidation of Calabresi; the distant and mysterious death of the fascist Nardi,[119] accused of assassinating Calabresi; the "suicide" of a good number of SID officers; the "accidental" deaths of several people who testified at the Piazza Fontana trial; the spectacular and simultaneous attacks against Coco and Occorsio [in June 1976], which were claimed – with the concern for symmetry that is always present in the spectacle of "opposing extremisms" – by both the Red Brigades and the fascists. It is thus worth remarking that both of these magistrates were more than a little involved in [the spectacle of] terrorism: Coco in the troubled and incongruous kidnapping of [Judge] Sossi; and Occorsio in the great spectacle of the prosecution of "the human beast," Pietro Valpreda. Naturally, all the mendacious sources of information presented (as confirmations of the official versions of these events) facts *that precisely contradicted them*: Coco "did not give in" to the RBs and *therefore* they took revenge against him, even though one doesn't understand why they didn't take their revenge by killing Sossi. I

---

[119] Gianni Nardi (1946-1976) alleged died in a car accident in Spain on 10 September 1976.

take a hostage, and I blackmail you; if you do not give in, it is you whom I kill, not the hostage?! Illogical logic, but spectacular logic, just the same.

As for Occorsio, he spent his final hours investigating the fascists; *therefore* they are the ones who had an interest in killing him. But, for heaven's sake, let no one advance the least suspicion about this logic. Namely: if Occorsio was occupied with the fascists during his final hours, after being occupied with the anarchists, but with equally poor results, then *someone* suggested to him to make the switch so that the fascists could be made to claim responsibility for his death, thus giving it an explanation (one couldn't accuse Valpreda of killing Occorsio *as well as* perpetrating the attack at the Piazza Fontana: he is a "guilty party" who is worn out, burned and unusable; if one were to read tomorrow that he had killed his mother-in-law, no one in Italy would believe it).

The judges who are currently occupied with the Moro affair are the least enviable people in Italy, and they must pay very good attention. They must take care to not lose themselves in their investigations or displease certain sectors of power; they must pay attention *to everything, always* because, for the State, the first opportunity to get rid of them will be the best one; and the RBs will soon after "claim" [responsibility for] their deaths, which will thus be explained to public opinion. And in Italy today, anything that can be explained *is thereby justified* and, if the explanation is improper (because no one can reply to it), it is an explanation that *cannot be appealed,* a lie that cannot be refuted and thus is no longer a lie. To speak as Ronchey does: if someone refutes it, it isn't refuted; if it is refuted, the refutation is not "believable"; if it is not "believable," *the refutation doesn't exist.* Few things that Orwell predicted in *1984* have not been verified. For example, read the following passage.[120]

> In some ways she was more acute than Winston, and far less susceptible to Party propaganda. Once he happened in some connection to mention the war (...) she startled him by saying casually that in her opinion the war was not happening. The rocket bombs which fell daily on London were probably fired by the Government (...) itself, "just to keep people frightened." This was an idea that literally had never occurred to him.

Several extra-parliamentarians, lost within their puerile illusions and

---

[120] George Orwell, Part Two, Section V, *1984,* quoted from the original English.

fetishistic ideology of armed struggle, would perhaps object that, since they *believe* in the armed struggle, other people, more "extreme" than they are, *could actually practice it* and be responsible for everything, including the kidnapping of Moro. I would respond that I have never doubted, either in public or in private, the imbecility of our extra-parliamentarians as a whole; but it is fitting to observe that, where they are concerned, they never doubt what the spectacle says about armed struggle or they themselves. Brave, alienated militants, pay attention to this only: if Moro had indeed been kidnapped and killed by free and autonomous revolutionaries, as the State has told you and as you believe, then it also follows that, *for the first time in ten years,* the State *hasn't lied about a matter concerning terrorism.* But this, being unbelievable and absurd, can be excluded.

On the contrary, the sad truth is that you *have always believed* all the lies of the past concerning Valpreda, Feltrinelli, the RBs and the rest. Even the anarchists' official newspaper, *Umanita Nova,* hastened to protect itself in the wake of the [attack at the] Piazza Fontana by separating its "responsibilities" from those of Valpreda, thus proving a courage suitable for its intelligence.

Many extreme-Left militants believe that they are shrewd because they understand that Pinelli didn't fall on his own from the fourth floor of the Central Police Station, but they will never manage to surpass their *record* for perversity, since they shortly thereafter applauded our secret services when they killed Commissioner Calabresi. Our bourgeoisie and Stalinists, who have already proved their inaptitude so well, thus have reasons to be consoled when they consider the stupidity of all their allegedly "extreme" adversaries, who in a certain way compensate for their own stupidity, even if it doesn't annul it. And indeed, in ten years, no extra-parliamentary groupuscule has ever managed to harm the State in the least, because none of them have been able to help the practical struggles of the wildcat workers in any way or to contribute to the advancement of their theoretical consciousness.

Impotent and maladroit, militants today accuse the State of being morally "responsible" for Moro's death *because it didn't save him* (and not because it was the one who killed him), just as in 1970 they accused the State of "moral responsibility" for the massacre at the Piazza Fontana, certainly not for ordering it done, but for not ordering the arrest of several fascists who were implicated in the affair, at least on the legal plane. The [newly elected] politicians who please themselves by imitating the gestures of the established ones continue to ignore the fact that politics has nothing to do with morality, but, rather, with the ideology that justifies certain policies, that is to say, all the lies that all politics normally require. This is why they always and only speak of the "moral responsibility" of the State and thus become re-responsible for all of its lies.

But let's reason absurdly; let's try for a single instant to consider the idea that the kidnapping of Moro was conceived and pulled off by subversives.[121] In such case, there would be several questions to be asked, and these are precisely the only questions that the contemplative militants have never asked, occupied as they are with admiring all that they are incapable of or with disagreeing with everything in which they do not participate. That is to say, *everything*.

Above all, one must wonder how it is possible that, over the course of two months, these subversives weren't able to accuse Moro of anything other than serving the interests of the bourgeoisie, instead of those of the proletariat, as if this was particular to Moro, as if there was no one in Parliament who was also "guilty" of this "crime"! The absurdity of such an accusation renders it perfectly unbelievable. Moro had never claimed or tried to make people believe that he defended the interests of the workers, despite what is said by the Stalinists or the extra-parliamentarians. To accuse him of such a "crime" is to accuse the rich of not being poor, or an enemy of not being

---

[121] Once again, we are compelled draw the reader's attention to the letter Sanguinetti wrote to Guy Debord on 1 June 1978: "Thus, the rest of the story of Moro and his death has led me to not exclude any hypothesis. And although what you wrote to me is completely probable and rational, [and though] it is as true as what I had thought, I will try here to envision this story in an inverted perspective: you will see that everything truly is possible (...) Thus, here is my reasoning and my hypothesis (...) The Italian Leftists are very stupid, obviously. But this same stupidity, on the one hand, isn't completely sufficient to render them all incapable of doing something and, on the other hand, is quite sufficient to convince them that terrorism can be a good thing. And you know that the Italian Leftist, unlike the French one, isn't a contemplator of theory, but a contemplator of practice (...) In fact, the same stupidity that had for a long time prevented them from understanding *from whence* came the attack of 1969 could very well have subsequently worked – when its provenance became *confusedly* clear to them – to make them 'theorize' that one responds to State terrorism with 'proletarian' terrorism. It is an unquestionable fact that there are many Leftists in Italy who have become terrorists in the last few years, and among them there are quite a few young workers (there are a hundred known groups). It remains to be seen if a similar blow is beyond their reach or not."

an ally. If these hypothetical "subversives" staged the "trial" of Moro to make such an accusation against him, they could have spared themselves the trouble and killed him along with his police escort on the Via Fani. But, as I have already said, behind this accusation was the opposite one. The abductors of Moro actually accused him of not sufficiently serving the interests of the bourgeoisie, and certainly not serving them too well.

Besides, the maladroit parody of "proletarian justice" clumsily staged by Moro's jailers *didn't even try* to get him to spit out the truth about the massacre at the Piazza Fontana or a hundred other, equally scandalous facts that any powerful man naturally knows, facts that would be highly instructive to the proletariat. Where this is concerned, we must remark that, if Moro in one of his first letters feared he would have to speak about "unpleasant and dangerous" truths, this did not worry anyone in the government, which shows that our ministers never feared anything of the kind, because *they knew* that they had nothing to fear. In their own proclamations, Moro's abductors never knew how or even wanted to address themselves to the workers, to whom they had nothing interesting to say. After having stated with assurance, just after the kidnapping, that "nothing will be hidden from the people," Moro's jailers, through his mediation, immediately began a long and secret correspondence with all the powerful men in the Christian Democratic Party, to whom the attack was a *warning,* and the kidnapping lasted for as long as was necessary to convince them all. The first *proof* of their convictions that they had to give was precisely that of not "negotiating" and they all hastened to give it. The terms for the release of the hostage – whom the RBs would have freed, at least officially, if the State agreed to release fifteen imprisoned militants – seem to have been set *only to be rejected,* certainly not because they were unacceptable to the State, but because (not being of any interest to any sector of the proletariat) these terms could not claim the support of any spontaneous or violent movement in the country – and Moro's jailers did not even try to inspire any such movement. Where the abductors betrayed their identities as agents of power, and did so in the most maladroit manner, was in their strong desire to be officially recognized by the existing powers: everyone from the ICP to the Christian Democrats, from the Pope to Waldheim.[122] This fact, and it alone, admirably proved that, not only did the abductors recognize the legitimacy of these powers, but also they were only preoccupied with being recognized *by them,* and certainly not by the proletariat. For their part,

---

[122] Kurt Josef Waldheim (1918-2007) was the Secretary General of the United Nations from 1972 to 1981.

the party leaders betrayed themselves when they admitted that the goal of the kidnapping was to divide the political forces of the government and added that it had failed to do so, when in fact it *succeeded.* The Christian Democrats and the Craxians quickly understood that they had to separate themselves, gently but resolutely, from the Stalinists. If Moro's jailers had been subversives, such a division would not have interested them, because any subversive knows that the only division likely to create disorder is the one made between the exploited and the exploiters, and certainly not any division between the different political parties that, in the spectacle, only represent the different forces that are used to maintain the same exploitation, even when the beneficiaries of it change. Finally, if Moro's abductors had been subversives, they certainly would not have given up the opportunity to release him, because Moro – calumnied by all his "friends" and betrayed by his recent allies – would have openly fought against all those whom he had previously protected. On the contrary, by killing him, the artisans of the Via Fani operation opportunely got out of difficulty all the powers, especially the Christian Democrats, for whom Moro was useful dead, but very harmful alive.

In any case, if Moro's abductors had been subversives, they certainly would not have chosen the freedom of Curcio and the others as the terms in the negotiations, because such terms would have given an excellent pretext for power to send the RBs packing and thereby not "lose honor." If his abductors were going to choose to pose unacceptable terms, they should have demanded *something other* than the liberation of only fifteen prisoners, and those who set unacceptable terms are always attentive to the fact that they should not be too easily rejected, as was the demand concerning those few brigadists. But in reality, Moro's abductor's did not want any of what they officially demanded: they knew that they could not openly demand what they really wanted without unmasking themselves. Today, *they have obtained* what they wanted. And, shortly before Moro's jailers did away with him, all the real terms of the blackmail *were inverted* with respect to the spectacular and official terms set for the Christian Democrats, and those real terms became this: either you change your policy or we will free Moro and you will see that *it will be him* who changes it. Things being what they were, the Christian Democrats and the "Socialist" leaders wisely preferred that *they be the ones* to change the policy at the expense of Moro, instead of risking a situation in which Moro changed it *at their expense.* Thus goes the world, despite all the flapping of the wings of the Capitoline geese, who claimed the opposite.

All of our incapable extra-parliamentarians, dazzled like primitive peoples by the technical success of the Via Fani operation, were not able to see beyond it by realizing that those who disposed of so many means and tactical capabilities surely would not put them at the service of a strategy that was as poor and unbelievable as the one attributed to the RBs, but, rather, at the service of a political design of much greater scope. But the extra-parliamentarians, faced with the operational efficiency on display at the Via Fani and in what followed it, naturally preferred to attribute it to "comrades who were mistaken" and not to enemies *who do not make mistakes* and calmly fuck people over. Here as well, our poor Leftists have taken their poor desires for reality, without suspecting that reality always surpasses their desires, but not in the manner that they desire. And if they were less ignorant, they would not have neglected the abilities of the unofficial Italian services so much and so wrongly. For example, they would know that, for Italy, the only war operations that were truly successful in World War II were the *commando* raids carried out by the Navy. It seems to me hardly necessary to recall how this brilliant tradition was admirably transmitted from the Navy to the secret services, first by Admiral Henke, who has never been an imbecile, and then by Admiral Casardi,[123] who is even more capable. Between them came the ignominious interregnum of Vito Miceli, an unskilled general who has in fact succumbed to his own ineptitude and Andreotti's prudence, which was not late in perceiving it. In fact, Andreotti did not have General Miceli arrested because he was responsible for the SID's "deviations" – which began well before Miceli's tenure, as Andreotti knows – but because Miceli's clumsiness threatened to blow the lid off the secret services' pot. Once more, Andreotti showed himself to be a finer politician than he wants to appear: he passed off his attack on Miceli as a concern for loyalty to the Constitution and thus won the predictable sympathies of a part of the Left. Andreotti's only error was, as usual, his false modesty and vanity. He rejoiced too much after Miceli's arrest, tried too much to appear naïve and declared on several occasions that, due to prudence, he had never wanted to occupy himself with the secret services, which was a scandalous declaration for a government leader, but necessary for someone who – having been occupied with them [out of necessity] – saw "things of which it is well to say nothing,"[124] things that were *so scandalous* that one could only keep quiet about *by feigning to not know about them*. And Andreotti knows that the scandal of ignorance is the price he must pay to feign ignorance of certain scandals. Nevertheless, he remains comical, like the fable

---

[123] Rear Admiral Ferdinando Casardi was the commander of the Italian 2nd Cruiser Division during the Battle of Cape Spada on 19 July 1940.
[124] Dante, *Inferno*, IV, 104.

in which the fox disguises himself as a lamb so as to be better accepted by the wolves.

Setting aside the admirals, we must note that, in Italy, there are excellent superior officers among the carabinieri, because not everyone is like Miceli or La Bruna[125] and only the Micelis and La Brunas get caught in the trap. On the other hand, there is a deeper and more dialectical argument that works in favor of the leadership of our secret services: if this era demands that certain men practice terrorism, it is also capable of *creating* the men who are needed by terrorism. Furthermore, one need not believe that the Via Fani operation was a super-human masterpiece of operational abilities. Just yesterday, even Idi Amin Dada[126] could pull off certain technical successes, about which the poor militants of Lotta continua will never cease to be astonished.

A great number of workers, many of whom I have encountered in the most diverse situations and who are much less naïve than the extra-parliamentarians, immediately concluded that "*they* kidnapped Aldo Moro," and by this they naturally meant those who have power. And to think that as recently as yesterday such workers voted and generally voted for the ICP!

The irreparable split that exists in this country between all those who have the floor (the politicians, the powerful and their servants, some of whom are journalists), on the one hand, and those who are deprived of the opportunity to speak, on the other, expresses itself perfectly in the fact that the former – who are far from the ordinary people and protected by the barrier of their bodyguards – *no longer know* what the latter say and think in the streets, restaurants and workplaces. As a result, the lies of power have become tangential; they have entered into a kind of autonomous orbit due to centrifugal force. And this orbit *no longer touches* any part of the "real country," in which the truth makes its way so much more easily because no obstacle hinders or intimidates it. In contrast, the spectacle has *become autistic*, that is to say, it is suffering from a schizophrenic psychopathological syndrome in which the ideas and actions of the sick person can no longer be modified by reality, from which he or she is irremediably separated and is thus constrained to live in his or her own world *beyond the real one*. Like King Oedipus, the spectacle has gouged its eyes out and continues blindly in its own

---

[125] Captain Antonino La Bruna worked in the domestic security branch of the *Servizio Informazione Difesa*. His superior officer was General Maletti.
[126] Idi Amin Dada (circa 1925-2003) was the military dictator and President of Uganda between 1971 and 1979.

terrorist delirium. Like King Oedipus, it no longer wants to see reality and, like President Andreotti, it says that it wants to know nothing about the secret services; it even proclaims that they were dismantled several years ago and no longer exist. If, like King Oedipus, the spectacle no longer wants to see reality, this is because it only wants to *be seen*, contemplated, admired and accepted as everything that it pretends to be. Thus, it wants to be heard, without ever hearing, and it even doesn't worry too much about not being heard. What seems to be the most important thing to the spectacle is that it can pursue its endless paranoiac voyage [undisturbed]. At the very moment that the police claim to make history, all historical facts are explained by power in a *police-related* way. The Hungarian psychiatric researcher Joseph Gabel[127] says that, according to what he defines as the "police conception of history," history is no longer constituted "by the entirety of objective forces, but by good or bad individual actions"; every event "is placed under the rubric of miracle or catastrophe." The interpretation of an event no longer involves its historical explanation, but the determination of its cause by either red or black magic. Thus, for power, the bombing of the Piazza Fontana was the *miracle* that made the unions renounce strikes and allowed the State to avoid civil war. In contrast, the death of Moro announced a mysterious *catastrophe* that, thanks to the skill and inflexibility of our politicians, spared us. But this has no importance to the large number of "plebian people" – to make use of an expression favored by the Stalinist Amendola – who have said, "If they kill Moro, it doesn't interest me at all: that's *their* affair," which is something I've heard thousands of times. "The country resisted; it knew how to react." What a good joke! The only reaction from this "mythological" country was (quite wisely) *to not believe* anything that one said to it.

Parallel to the catastrophic or miraculous explication of history, the spectacle comes to *no longer know* what it dominates, no longer grasps hold of the reality and thoughts that it urgently must master. As Machiavelli says, "when one knows the least, one has the most suspicions." The entire population, and the young people in particular, become suspect in the eyes of power. At the same time, if artificial terrorism claims to be *the only real phenomenon,* all the spontaneous revolts – such as those in Rome and Bologna in 1977 – become, in accordance with the "police conception of history," *a conspiracy* that has been artificially plotted and conducted by forces that are

---

[127] Joseph Gabel (1912-2004) was a Hungarian-born French sociologist and philosopher. His book *La Fausse Conscience: essai sur la reification* was first published in 1962.

"hidden" and yet "quite identifiable," which is what the Stalinists believe even today. Everything that power cannot foresee, because it hasn't organized it, thus becomes a "conspiracy" against it. On the other hand, artificial terrorism, since it is organized and conducted by the masters of the spectacle, is a *real* and spontaneous phenomenon that these masters continually feign to combat for the simple reason that *it is easier to defend oneself against a simulated enemy than a real one.* And power would like to refuse the very *status* of enemy to its real enemy, which is the proletariat. If the workers say they are against this demented terrorism, "they are for the State," and if they are against the State, "they are terrorists," that is to say, enemies of the common good and thus *public enemies.* And against a public enemy, everything is permitted, everything is authorized.

Gabel goes on to say that "the police conception of history represents the most extreme form of political alienation (...): unfavorable events can only be explained by external actions (the conspiracy) and they are experienced (by the sick person) as an unexpected, 'unmerited' catastrophe." And this is why any spontaneous strike must be an insult to "the working class," which is so well represented by the unions, and any wildcat struggle is "provocative," "corporative," "unjust" and "unmerited." All this goes back to the clinical framework of autistic schizophrenia. "The *syndrome of external action* (...) is the clinical expression for the irruption of the dialectic in a reified world that can only accept the event as a catastrophe" (J. Gabel, *False Consciousness*). The irruption of the dialectic corresponds to nothing other than the *irruption of struggle* in a reified world, which, more exactly, is the spectacular-commodity world, which cannot accept struggle, *even in the domain of thought.* Thus, this spectacular society is no longer even capable of *thinking.* Those who reason logically, for example, can only accept the identity of two things when it is based on the identity of subjects. In contrast, the spectacle, which is para-logical, establishes identity on the basis of the identity of predicates and thus says: "the devil is black; that which is black is the devil," or "the Jew is bad; that which is bad is Jewish," or "terrorism is catastrophic, the catastrophe is terrorism." Aside from terrorism, everything else goes well. Unfortunately, there is terrorism: what can we do about it?

If I say, "a police officer must have a legally unblemished record; Mario Bianchi[128] is a police officer; therefore he has a legally unblemished record," the schizophrenic will say, "Mario Bianchi has a legally unblemished record, *therefore* he is a police officer." Thus the spectacle, when it has reached the

---

[128] Mario Bianchi (born 1939), an Italian filmmaker.

point of autism, says, "those who kidnapped Moro are terrorists; the RBs are terrorists; Moro was kidnapped by the RBs." No identification is improper to the spectacle, except for one, which is the only one not made. Namely: the State has proclaimed for years that it is combating the RBs; it has infiltrated them several times without ever trying to dismantle them; *thus the State makes use of the RBs* as a cover, because *the RBs are useful to the State,* thus RBs = the State. Power has confessed in a thousand different ways that it fears the making of such an identification: for example, when it invented the neurotic and maladroit slogan, "Either with the State or the RBs," which means "Either with me, or with me."

A long time before the advent of the spectacle, religion – which has always been a functioning ideological prototype for all the old forms of power – had already invented the Devil, the first and supreme *agent provocateur,* whose role was to assure the complete triumph of the Kingdom of God. Religion projected the simple necessity of concrete and real power upon the metaphysical world. Thus Cicero *needed* to amplify the risk constituted by Catilina to magnify his own glory as savior of the fatherland and to multiply his own abuses in this way.[129] For any power, the only real catastrophe is being swept from history, and each power, once it has become weak and senses the imminence of this real catastrophe, has always tried to consolidate itself by feigning to fight an unequal battle against a convenient adversary. But such battles have also been the final prayers *for hearth and home*[130] made by power in trouble. History is full of examples. According to Paul-Louis Courier,[131]

> As scandal is necessary for the greater glory of God, conspiracies are necessary for the maintenance of the political police. They produce them, smother them, change their look, reveal them – this is the great art of government ministers, the strong point and goal of the science of statesmen, the *transcendental politics* that we have recently perfected [here in France] and that the jealous English want to imitate and infringe, but crudely (...). From the moment that one knows what they want to do, the ministers cannot do it or no longer want to do it. Politics known is politics lost; affairs of State, State secrets (...). Decency is necessary for a

---

[129] Cicero (106-43 BCE) was the Consul during the conspiracy organized by Lucius Sergius Catilina.

[130] Latin in original.

[131] *Author's note*: Paul-Louis Courier, *Pamphlets politiques.*

constitutional government.

Courier wrote this in 1820, at the height of the Restoration. Today, fearing a new and more frightening revolution, the State uses the same practices as before, this time at a much higher level, to obtain a *preventive* Restoration. The "transcendental politics" of the past is the imminent politics of the spectacle, which always presents itself as "the adversary of all evil," as Dante said of God; thus, according to the spectacle's autistic logic, all that is opposed to the spectacle itself *is evil.* And faced with this pitiless, preventive Restoration – this despicable series of provocations, massacres, assassinations and lies that seek to camouflage a crystal-clear reality – there are growing numbers of sociological "studies" and enslaved and progressive journalists who, having a better grasp on their [financial] security than on the simple reality of the facts, compete with each other in their expressions of a "certain sympathy" for "armed struggle" and clandestinity, as does the unspeakable Giorgio Bocca under the pretext that all this reminds him of his epic struggles as part of the Resistance. Men like Bocca are, so to speak, "legitimate" when, under the influence of fear, they declare that they feel sympathy for terrorism, because they earn four or five million a month, and because they feel that the existence of terrorism guarantees them that this income will continue. But those who have nothing are *deceived* by these men, who always lie with the goal of perpetrating their dirty tricks easily and at the expense of others. People like you, Bocca: one doesn't kill them; that would be showing them too much respect! No one wants to see you die, but, for my part, if I ever encounter you on the street, be certain that *I will teach you how to live,* you idiot.

On the other hand, there is the attorney Giannino Guiso, who tells us about the ideological subtleties of Curcio, the sociologist Sabino Acquaviva, who expands upon grandiloquent "explanations" of terrorism, and the pedant [Mario] Scialoja, a journalist for *L'Espresso,* who pretentiously discourses about the "strategies of armed struggle"; and all of them feign to be "in the know" about the secret affairs of the social revolution by seeking to give credibility to artificial terrorism as a prelude to revolution. But

You will be surprised, when you come to the end,
That you have not persuaded us of anything.[132]

---

[132] *Note by Jean-François Martos*: A quotation from a work by Paul-Louis Courier, in French in the original Italian version of this text. Note that this quotation slightly modifies a line in Moliere's *L'Ecole des femmes* by replacing

Gianfranco Sanguinetti

Respected hoaxers, I have only one thing to say to you: unlike you, for the last thirteen years I have known a large number of the revolutionaries in Europe – they are also known by all the police forces – who have contributed the most, in both theory and practice, to reducing capitalism to its current conditions and, without exception, *none of them* have ever practiced or even applauded modern spectacular terrorism, which is a fact that appears obvious to me. *There are no secret affairs of the revolution*: today, everything that is secret belongs to power, that is to say, to the counter-revolution. And all the police forces know this perfectly well.

Gentlemen of the government, it is fitting that, from now on, you have a calm conscience on this point: as long as your State exists, and as long as I am alive, I will never stop denouncing the terrorism perpetrated by your unofficial services, whatever the costs, because doing so is *the primary concern of the proletariat and the social revolution* at this moment in this country. And this precisely because, as Courier says, "known politics is politics lost." And if this criminal State continues to lie, kill and provoke the entire population, it will henceforth be constrained to take off its "democratic" mask, act against the workers in its own name, abandon the current comedic spectacle in which the secret services display themselves (thereby supporting the illusions of naïve militants about the "armed struggle," which are in turn used to render those services' provocations plausible), and throw into prison hundreds of people, while the police forces train themselves for civil war by shooting at sitting ducks.

Ever since 1969, the spectacle, to continue to be believed, has had to attribute *unbelievable* actions to its enemies, and, to continue to be accepted, it has had to ascribe *unacceptable* behavior to proletarians. As a result, the spectacle has generated enough publicity that the people who allow themselves to be frightened will choose "the lesser of two evils," that is to say, the current state of things. When *the real leaders* of the RBs ordered that unarmed people be shot in the legs – something that is only worthy of police-like cowardice and certainly not worthy of revolutionary courage – and when those leaders ordered such attacks, which struck second-tier industrial executives, they knew exactly what they wanted to accomplish, which was to frighten that part of the bourgeoisie that doesn't have sufficient class consciousness (because it doesn't enjoy the advantages of the big bourgeoisie) and to win it over to the side of the latter with the upcoming civil war in mind.

"persuaded me" with "persuaded us."

The fragility of such artificial terrorism lies in the fact that, when one adopts such a tactic, it becomes known and thus judged; as a result, everything that gave this tactic its force now weakens it, and thus the great advantages that it assured its strategists become a major inconvenience.

The current President of the Republic, Pertini,[133] who is a naïve man, always and only fears fascism, because he only fears what he knows. But from now on, what he must do is fear *what he doesn't know* and *know what he must fear today as quickly as possible,* that is to say, not an overt dictatorship, but a formidable, hidden despotism of the secret services, one that is all the more powerful because it uses its strength to vigorously affirm that it doesn't exist. It is not at all by chance that, in September 1978, Fanfani [almost] imperceptibly invented a new and important cabinet post that has no precedent in our institutional history: Advisor to the President of the Republic for Problems of Democratic Order and Security. And it was not at all by chance that, to fill this position, Fanfani called upon Major General Arnaldo Ferrara, who is considered – where military matters are concerned – to be one of the best officers in the carabinieri and Europe as a whole. By putting on old Pertini's side a young general like Ferrara, "a man with eyes of ice and refined tastes," Fanfani has not only institutionalized a *de facto situation* by sanctioning the power attained by the unofficial services, but he has also taken the first step towards crowning his old dream of a presidential republic. Arnaldo Ferrara, an intelligent and refined officer, who recently refused to become the head of the SISDE (the secret service attached to the Minister of the Interior) and who didn't give in to Andreotti's insistence that he renounce his own personal ambitions – this superior officer who "has penetrated into the most secret of the State's secrets and those of the men who represent it" (as Roberto Fabiani has assured us) – is in fact *the new President of the Republic.* Moreover, Ferrara now possesses powers that no President of Italy has ever had. In fact, his position as "advisor" (an honorific title in appearance only) guarantees him more and better powers than any other official and, at the same time, a freedom of action whose limits are difficult to determine but easy to surpass. Faced with such developments, the proletariat can only combat them on open ground or get used to them by tolerating of all their serious consequences.

If one truly wants to know it, this is the precise purpose of outfitting the Presidency of the Republic with a man "beyond all suspicion": it serves to hide the Republic's *end* and its "painless" transformation into a police State, all the

---

[133] Sandro Pertini (1896-1990), a member of the Italian Socialist Party.

while maintaining the spectacle of "democratic" appearances. The Honorable Pertini – since he has always remained at the margins of his own political party and, as he is the only politician who (never having had real power before) has always been a stranger to the practices of the unofficial services – is thus the man who *knows the least of these practices* and who offers the best qualities to be manipulated by hidden powers without realizing it. The detached units of the State, having reached their current level of power, can only continue to make use of the same tactics of *infiltration* that were used with success on the RBs, but this time they will be extended *to all of the State's institutions.* In these conditions, not only will terrorism not cease, but it will *grow* quantitatively and qualitatively;[134] and one can already foresee that, if a social revolution does not put an end to this tragic farce, Pertini's presidency will be *the most dire* period in the history of the Republic. And so that someone doesn't come to tell me that what I say is "very serious": I know that perfectly well, but I also know that to keep quiet, as all the others do, is *even more serious,* and that the most serious phenomenon is the one that everyone witnesses without ever denouncing. There is nothing secret in this phenomenon, which nevertheless remains *undisclosed* to the general awareness and, as Bernard Shaw said, "there are no better kept secrets than the secrets that everyone guesses."[135] And consciousness *always comes too late.*

In such conditions, the first duty of all conscious subversives is to *pitilessly chase all illusions about terrorism from the heads of those called to action.* As I have already said elsewhere, historically speaking, terrorism has never had any revolutionary effectiveness, except when all other forms of subversive activity have been rendered impossible by complete repression and an important part of the proletarian population has been led to take part in terrorism silently.[136] But this is no longer or *still not* the case in

---

[134] Sanguinetti was quickly proven right. On 2 August 1980, the central train station in Bologna was the site of the worst terrorist attack in Italy to date. It was immediately (and falsely) blamed upon a neo-fascist group, the *Nuclei Armati Rivoluzionari.*

[135] George Bernard Shaw, Act III, *Mrs. Warren's Profession* (1893).

[136] *Author's note*: Cf. the manifesto entitled *Benvenuti nella citta piu libera del mondo* ["Welcome to the Freest City in the World"] and distributed in Bologna, Rome and Milan on 23 September 1977. [*Translator*: it seems to me that this is a very important but unacknowledged modification of the sweeping claims made at the beginning at this text, namely, "The attacks by the Palestinians

contemporary Italy. Moreover, it is fitting to note that the revolutionary effectiveness of terrorism has always been *very limited,* as the history of the end of the 19th century has shown.

In contrast, the bourgeoisie, which established its domination in France in 1793 thanks to terrorism, must have renewed recourse to this weapon (in a strategically defensive context) during a historical period in which its power is universally being placed in question by the very proletarian forces that its own development has created. At the same time, the bourgeois State's secret services cover for their terrorism by making opportune use of the most naïve militants of a Leninism that has been completely frustrated by history, a Leninism that, between 1918 and 1921, also used the same anti-worker terrorist methods to destroy the soviets and seize control of the State and the capitalist economy in Russia.

All States *have always been terroristic,* but they are more violently so during their births and when they face the imminence of their deaths. And those today who, either due to despair or because they are victims of the propaganda that the regime creates in favor of terrorism as the *best example*[137] of subversion, and who thus contemplate artificial terrorism with an uncritical admiration (and even try to practice it on occasion), do not know that they are only *competing* with the State *on its own terrain* and that, on this terrain, not only is the State stronger, but *it will always have the last word.* Everything that does not destroy the spectacle *reinforces it,* and the incredible reinforcement of all the governmental powers of control that has taken place thanks to the pretext of [fighting again] spectacular terrorism has already been used against *the entire Italian proletarian movement,* which is the most advanced and most radical in Europe today.

For us, it is certainly not a question of "disagreeing" with terrorism in a stupid and abstract manner, as do the militants of Lotta continua, nor is it a question of admiring the "comrades who are mistaken," as do the so-called Autonomes[138] (who thereby give the despicable Stalinists a pretext to preach in favor of informing on others systematically). On the contrary, it is a question of simply judging terrorism according to its actual results, who

---

and the Irish, for example, are acts of *offensive* terrorism," and "Experience has long since shown that, if they are part of a strategic offensive, they are always doomed to failure."]

[137] Latin in original.

[138] Who have been discussed in this text under the rubric of the "extra-parliamentarians."

practices it and what usage the spectacle makes of it, and finally coming to *conclusions* about it.

The true terrorism is the practice of continually obligating everyone to take positions for or against mysterious and obscure events that are prefabricated with this precise intention in mind. Furthermore, continually constraining the entire working class to come out against this or that attack, to which everyone except the unofficial services of the State are strangers, is what permits the union bureaucrats to unite under their anti-worker directives the workers of every factory in turmoil, where some mid-level executive is regularly shot in the leg.

In 1921, in the midst of the repression of the Kronstadt soviet, when Lenin famously declared "here or there with a gun, but not with the opposition of the workers; we have had enough of the opposition of the workers," he showed himself to be less dishonest than Berlinguer, who said, "with the State or with the RBs," because he had no fear of declaring that his only goal is the liquidation of the opposition of the workers. Well, from the precise moment that someone affirms that he or she is "with the State," *he or she knows that he or she supports terrorism,* which, in this case, is the most putrid State terrorism that has ever been deployed against the proletariat. Such a person knows that he or she supports those responsible for the deaths at the Piazza Fontana, on board the *Italicus,* and at Brescia, as well as the assassins of Pinelli and a hundred other people. Such a person should no longer break our balls because we have had enough of the crocodile tears shed for the "martyrs" of the Via Fani and enough of the provocations, the crude efforts at intimidation, the assassinations, the prison sentences, the brazen hypocrisy of the defense of the "democratic institutions" and all the rest.

As for us, the subversives, who support the opposition of the workers and do not support the State, we will prove ourselves to be so, above all and on every occasion, by continually unmasking all the acts of terrorism perpetrated by the secret services of the State, to which we willingly leave *the monopoly on terror,* and by making the State's infamy more infamous by publicizing it: by giving it the publicity that *it merits.*

When our turn comes, we won't be lacking weapons or valorous fighters. We are not slaves to the commodity fetish of weapons, and *we will procure them* when they are necessary and in the simplest fashion: by taking them from your generals, police officers and bourgeois, because they already have enough of them for all the workers in Italy. "We do not have compassion [for you]; we do not expect any from you. When our turn comes, we will not embellish the violence" (Marx).[139]

A thousand repetitions of [the attacks at] the Via Fani and the Piazza Fontana will not benefit capitalism as much as a single anti-bourgeois and anti-Stalinist wildcat strike or a simple act of sabotage against production hurts it. Every day, millions of oppressed minds wake up and revolt against exploitation, and wildcat workers know perfectly well that the social revolution does not make its way by accumulating dead bodies, which is a prerogative of Stalinist-bourgeois counter-revolution (a prerogative that no [true] revolutionary has ever contested).

As for those who have joined up with alienated and hierarchical militantism at the moment of its bankruptcy: they can only become subversives *on the condition that they leave militantism behind,* and only if they succeed in negating in acts the conditions (set by the spectacle itself) for what is today designated by the vague but just term "dissidence," which by nature is always powerless.

From now on, those in Italy who do not use all the intelligence that they have to *quickly* comprehend the truth that is hidden behind each lie told by the State are *allies* of the enemies of the proletariat. And those who still claim they want to combat alienation *with alienated means* – militantism and ideology – will quickly realize that they have renounced *real* combat. It will certainly not be the militants who will make the social revolution, nor will the secret services and the Stalinist police be able to prevent it!

---

[139] Karl Marx, "The Summary Suppression of the *Neue Rheinische Zeitung*" (1849). Note that most translations from the original German render the last line of this passage "When our time comes, we will not make excuses for the terror."

Press Clippings[140]

This Leftist *"Action Directe,"* my dear, "the French equivalent of the Italian Red Brigades," exploded bombs in the midst of the cleaning ladies at Orly Airport (...) Seven injured. You say it is Leftist. "Action Directe" fired upon police officers on guard in front of the Iranian embassy, injuring two people. Despicable Leftists! Unluckily, detectives from the *Brigade criminelle* arrested one of the truly guilty parties. He was a detective from the *Renseignements généraux,* also an extreme-Right militant and someone implicated in pro-Corsican-independence attacks (all Leftists, as well).

Complete silence has greeted Gianfranco Sanguinetti's book, *On Terrorism and the State,* which I recommended to you three weeks ago. Everyone should read it; that is why no one has spoken to you of it. Certain people in France need "Action Directe" to exist. Thus one has provoked the birth of the French Red Brigades; one has even fabricated them, because such a group can always be useful if power threatens to change hands. Those who push the passage to "direct action" are the very ones who have reasons to present people like Piperno and Pace as the assassins of Aldo Moro. Have you noticed their formidable discretion since the arrest of the detective from the RG? This cop was in possession of the weapon that fired upon the gendarmes; in other words, this should make waves. This same cop was tasked by the RG with surveilling anarchist groups. Snicker and meditate upon that.

(Delfeil de Ton [Henri Roussel], *Le Nouvel Observateur,* 5 July 1980.)

This implacable diagnosis comes from Milan and Gianfranco Sanguinetti, whose book has been quashed in Italy by the political police. Nevertheless, this importer, this smuggler, of "situationism" into Italy attained runaway sales with his "Truth Report on the Last Chances to Save Capitalism in Italy," signed Censor, behind whom everyone thought they saw a politician close to power (...)

Sanguinetti keeps it short and hits home. For him, *"those who want power in Italy must demonstrate that they can manage terrorism,"* and *"if the States have recourse to direct terrorism, they must perpetrate it against the population."* For him, there is no doubt that the [recent] attack in Bologna

---

[140] "Several Judgments by the Commentators on Gianfranco Sanguinetti's *On Terrorism and the State,*" collected and published as an insert to the second edition of *Du Terrorisme et de l'Etat* (Le fin mot de l'Histoire, 1981).

must immediately be connected with the one at the Piazza Fontana in Milan in 1969, where the role of certain Italian secret services has finally been exposed. Thus, it appears that *"the original terrorist group"* objectively becomes, not only an accomplice, but *"a defensive appendage of the State."* In fact, Sanguinetti says, they are the same people! Moro was assassinated to terminate the "historic compromise" between Communists and Christian Democrats, and also to warm "liberals" who are a little too tempted by timid reforms.

The text was written in April 1979. This past Saturday, 2 August [1980], Aldo Moro's widow, heard by the ad hoc investigatory commission, confirmed that, during an international summit, her husband was warned or, rather, threatened that he should change course as soon as possible or leave the political stage. She also emphasized how Andreotti, then the President of the Council, and the current one, Cossiga, who was then the Minister of the Interior, refused to give Moro an armored vehicle. Case made, and Moro's widow is not a "situationist" (...)

There will be other attacks, Sanguinetti easily predicts, because *"like a drug, artificial terrorism necessitates and demands to be administered in always more massive and more frequent dosages."* One comprehends that, by describing all this, he has hardly made friends. One comprehends that his book has hardly been discussed, because it insults almost all the journalists in Italy "who count" by treating them like lachrymatory sheep. And yet all he had to do was think logically to create this small bomb of probability about Italy and the modern State.

(Dominique Durand, *Le Canard enchaîné,* 6 August 1980.)

The publication of the "Proofs of the Nonexistence of Censor by His Creator," which explained the hoax that had victimized the "well-informed minds" of Italy and openly exposed their misery, as well as instructions that come "from above," no doubt explain the silence in Italy that has fallen upon his most recent publication (...)

Even "Libération" [is silent]. This is not a question of "not dispiriting Billancourt," but of "not dispiriting Lotta continua."

The specialists in Italy at this newspaper have known about the book since April, but have said nothing about it. Here and there one grants that Sanguinetti "is a brilliant person," but F.L. "has read the summary and finds his theories absurd." Our prized Katia Kaup proposed to publish a few "good pages" the day of the funerals, but nothing came of it, though the profound reason for the hostility towards Sanguinetti's book was finally revealed. "*He has a house in the country and goes horse-riding.*" One will understand that I jumped at the occasion (...) to speak of this book, which seemed cursed until now (...)

More than a year before the new attack in Bologna, Sanguinetti announced in his book that, "*the detached units of the State, having reached their current power, can only continue to make use of the same tactic of infiltration used with success with respect to the RBs, by extending it today to all the institutions of the State. In these conditions, not only will terrorism not stop, but it will increase, qualitatively and quantitatively.*"

In fact, the author only recognizes a single [form of] terrorism: "the terrorism of the truth"; for him, "*a thousand Via Fanis and Piazza Fontanas will not benefit capitalism as much as it is harmed by a single anti-bourgeois and anti-Stalinist wildcat strike or a simple, violent and successful sabotage of production (...)*"

Gianfranco Sanguinetti has been prohibited from visiting France since 21 July '71.

(Hélène Hazera, *Libération*, 18 August 1980.)

"(...) the most intelligent book one has written on the question (...) tragically confirmed by the attack in Bologna (...). In my opinion, what he has said is greatly applicable to terrorism in France (including Corsica) (...) a book that we should study in our departmental groups."

(L'Ecole Emancipée, 5 September 1980.)

In the last ten years, there haven't been a dozen works, and I am being generous, that have been so completely subversive. With the result that we cannot give any review of it alongside the cultural chatter, which would be absolutely inappropriate. Get a copy now: although our French State is formally liberal (as well as cuntish), we do not see how a text as manifestly harmful can long remain freely available. But, on the other hand, its means of distribution (...) is such that its circulation cannot be suppressed.

(The Manchette Bros, *Charlie Hebdo*, 17 September 1980.)

Perhaps the strongest page in this very strong little book that Gianfranco Sanguinetti dedicates to so-called "Leftist" and "black" [fascist] terrorism in Italy is simply the reprint of a manifesto that he and his friends published in 1969, exactly one week after the massacre caused by the bombing of the Piazza Fontana in Milan.

Soon after, Valpreda, the "guilty party," was arrested and spent several years in prison before his complete innocence was recognized and the real assassins, who were neo-fascists, were put on trial. Meanwhile, unfortunate Pinelli, who got in the way, was defenestrated. Later, the shady police commissioner Calabresi was liquidated because he was too involved and knew too much about it. In 1969, all of the press and the official voices screamed about the "anarchist" attack. But right from the start, Sanguinetti and his friends stated that it was a staged attack, a provocation. Which today is obvious.

Whether it's Red Brigades or Black Brigades, the number of infiltrated agents in their ranks (from the monk Girotto to the informer Pisetta), the role that the Italian secret services have played (from the Borghese conspiracy and the "Rose des vents" to the arrest of General Miceli, head of the SID), and the strange discoveries produced by each new "red" or "black" attack (from the massacre at Brescia in 1974 to the massacres in Bologna in 1974 and 1980, and from the kidnapping and assassination of Moro to the most recent attacks) makes things such that, from this sorcerers' cauldron, comes the powerful scent of machination, an odor comparable to the stench that comes from the archives of the Czarist Okhrana – which, under the direction of the famous Azev, did not fear to have its own agents kill Plehvo, the Minister of the Interior, or Grand Duke Serge, the uncle of the Czar, and which had "raised" its "Lenin agent," Roman Malinovski, up to the Bolsheviks' Central Committee. In Rome today as in St. Petersburg yesterday, who profits from terrorism? The response is clear: not the people.

(Claude Roy, *Le Nouvel Observateur*, 6 October 1980.)

(...) with biting conviction and in an implacable style, stuffed full of Latin quotations and verses from Dante, he accuses the Italian secret services of having organized *all* the attacks (including the assassination of Aldo Moro) to reinforce the unsteady national consensus. *Id fecit cui prodest?*

(Guy Rossi-Landi, *Lire*, October 1980.)

A "scandalous" book (...) Two years later, nothing has contradicted Sanguinetti. And in any event, reading these 139 pages will much better inform the reader than the torrents of bad literature unleashed on the question ever since the spectacle of terrorism took hold of the mediatic stage.

(Marc Kravetz, *Magazine Littéraire,* January 1981.)

"Postface to the Dutch translation of Gianfranco Sanguinetti's book *On Terrorism and the State*"[141]

In Holland, in a region among the least impoverished, the most moderate and the most "democratized" in this poisoned world, where one can get together to criticize the quality of the heroin, and where pneumatic drills that have chased away the inhabitants are subsequently displayed, with the graffiti that denounces them, in the city's subways like works of art – here in Holland as well as elsewhere the taste to follow the excellent example of our Italian comrades grows: "their absenteeism; their wildcat strikes that no particular concession can appease; their lucid refusal of work; their scorn for the law and all the Statist political parties" (Guy Debord, *Preface to the Fourth Italian Edition of "The Society of the Spectacle"*). Here as elsewhere, the conditions that render life impossible for us force us to struggle – to engage in the only struggle in which it is still worth the difficulty of investing our talents and in which the possibilities of deploying these talents are infinite. Therefore, if we want to bring this struggle to a good end, it is necessary to know the enemy's weapons, and their uses, so as to turn them against it, or at least reduce those weapons to impotence.

Dutch commentators aren't more innocent than their Italian colleagues, but they are completely indifferent towards the truth. And it goes without saying that, among us as well, all politicians and union leaders lie to the same extent that the industrialists make profits by selling lime as insecticide, and insecticide as food, because, here and there, the truth serves them so little. Moreover, no one knows how to discern the truth any longer, with the exception of the comrades who think on the basis of a proletarian perspective and who have nothing to lose and everything to gain in it. It is for them that I have translated this book.

And to please these comrades even more, to provide all suitable clarity to the theses that are defended here with so much verve, but not always with as much precision, we originally intended to introduce them with Debord's *Preface*, as translated by Jaap Kloosterman. Sometimes Sanguinetti's book

---

[141] Written by Els van Daele, dated 1 May 1981, and published in *Over het terrorisme en de staat: de theorie en praktijk van het terrorisme voor het eerst wereldkundig gemaak* (Bussum: Wereldvenster, 1982). Translated into French as "Postface à la traduction hollandaise du livre de Gianfranco Sanguinetti, *Del Terrorismo e dello Stato*" and published in *Editions Champ Libre, Correspondance, Vol. 2* (Editions Champ Libre, Paris, 1981), pp. 118-124.

gives the impression that its author needs to persuade himself of the validity of his own theses, which the author of *The Society of the Spectacle* did not need to do. As there are a large number of confluences between these two books, from the choices of historical examples to certain stylistic details – from which one could deduce a close collaboration[142] – the pages of the *Preface* that deal with the same aspect of the class struggle, on the same terrain, and at the same time, might seem to the reader to be a summary of *On Terrorism,* but the same disturbances are in fact analyzed in it with a method and a rigor that are lacking in Sanguinetti's exposition. By contrast, the *Preface* lacks – and this is very good – the laborious and abstract schemas in which Sanguinetti believes he must and can classify *all terrorism.* By limiting himself to speaking of the maneuvers of the Red Brigades, in general, and the execution of Moro, in particular, "Gianfranco Sanguinetti shelters *On Terrorism and the State* from all critique (...) To speak of the RBs as an extension of the Italian secret services indeed no longer appears well founded," as a comrade in Paris has noted.[143] Without concerning himself with history, Sanguinetti banishes [from his analysis] the many forms of terrorism that, in our century alone, have been and are still employed, not only by the State or by the mafia, but also by the most implacable enemies of the State and political economy, as much offensively as defensively, as *one* weapon in the struggle.[144] By only

---

[142] *Author's note*: we recall that these two authors co-signed the principle text in *The Veritable Split* and that Debord translated Censor's *Truthful Report* [from the Italian into French].

[143] *Author's note*: *Rien qu'un pion sur l'échiquier,* anonymous tract published in Paris, February 1981.

[144] *Author's note*: Thus, in Spain, apart from the ETA and GRAPO (which fulfill exactly the same function as the RBs in Italy), one has seen at work many autonomous libertarian groups that do not fit at all into the categories of Sanguinetti's [concept of] terrorism, but that have, all the same, dynamited railroad lines and attacked businesses and banks. These groups have conceived of their actions in the much more fecund theoretical framework of *the armed struggle of the proletariat.* They have undertaken their operations as a part of, and as support for, the offensive strikes of Spanish workers, which especially marked the years 1976-1978. And these are groups that, in general, have taken up the most advanced [theoretical] positions. "One must not forget that the major part of the workers' movement still scorns theory, considering it to be the work of intellectuals. By contrast, we scorn the 'intellectuals' who don't have the passion to put revolutionary theory into practice, and never

implicating *State* terrorism in his critique (the ETA and the IRA want to conquer the State, while the RBs and GRAPO exist to defend it), and by presenting this critique as a *general* one, Sanguinetti – at the beginning of the 10th chapter of his *Remedy for Everything* – places all armed struggle in a bad light, and, by further developing several nuances, he only manages to contradict himself and to unintentionally demonstrate that his schema is defective. "This schema cannot be vaguely imputed to an error in judgment. It finds its truth in an active policy of wait-and-see ('I would consider myself *hardly practical . . .*') that Sanguinetti sets up as the *non plus ultra* of the revolutionary attitude that is not possessed by the 'bad workers' to whom his book is dedicated" (*Rien qu'on pion*). And yet the author loudly demands to be in the first position as *the* "specialist" in the denunciation of Italian State terrorism, today and in the future.

But it so happens that he was already not up to this pretention when he formulated it – because of what we can read in a letter written by Guy Debord to Jaap Kloosterman on 23 February 1981:

> After the end of our organizational links in 1972, for several years I maintained a very close collaboration with Gianfranco on several projects and very good personal relations [as well]. But all this is over. At the moment that Moro was kidnapped, I wrote to Gianfranco and revealed the truth of this entire affair, advised him to reveal it [in Italy] immediately and, at the same time, go underground, since he was, in any case, in great danger, because the enemy knew that – having written *Censor* – he was probably the only one in Italy who *could* possibly reveal this truth at that very moment, that is to say, when the enemy absolutely didn't want to run this risk, when Moro was still alive, etc. (To reveal what had taken place once the affair was over, almost forgotten, and other spectacles had taken the stage, would only express 'an opinion,' although a dangerous one, certainly.) For reasons that have remained very obscure to me, Gianfranco *then* responded that my thesis – which he subsequently took up – was brilliant

take up theory – which we make use of – against themselves. This is what we call theoretical expropriation" (see *Appels de la prison de Ségovia*, Paris, Champ Libre, November 1980). Before giving up the ghost, the last Spanish government in place before the military coup of 1981 was forced to free the guiltiest of these comrades, who were all in prison.

and ingenious, but *he* believed that it was true Leftists who then held Moro captive. Nevertheless, this was a belief that no slightly reasonable person, very up-to-date with the Italian situation until the day before these events, could entertain.

The idea that true Leftists had kidnapped Moro was a belief that no one in Holland alerted by *Censor* and having the occasion to read a few foreign newspapers could entertain, either.

And yet the author of *Censor,* who said to us on 16 March 1978 that he "has not been able to keep himself from thinking" that the kidnapping of Moro was the work of the Italian secret services, managed to prevent *other people* from subsequently choosing to reject this idea – and [so] once again the spectacle obtained its [desired] effect and succeeded in hiding the truth for as long as was necessary. The spectacle isn't only effective when it hides a secret or when one believes what it says; it is even more so when it is considered as an enigma to be resolved or when one doesn't know how to combat it. When Moro was kidnapped, Sanguinetti failed to intervene. And, in its turn, the fact of keeping his error hidden determined the course of all his subsequent actions. No doubt it was his bad conscience that dictated this promise to him: "As long as your State exists, and I am alive, I will never stop denouncing the terrorism of your parallel services, and no matter what," but *post festum*.

It is certainly not by keeping such secrets that one obtains the position of fundamental superiority from which one "can attack and successfully combat all the forces of thoughtlessness" [and] *vanquish* them. And it is not by passing over in silence the fact that *someone else* had known these things, and known them so well, that one prevents the revelation of a truth of which one is ashamed. But what cruel irony it is that this revelation took place due to the fact that Dutch comrades wanted to add Debord's *Preface* to Sanguinetti's *On Terrorism*[145] – the very *Preface* that Sanguinetti never mentioned, not even in the 1980 French edition of his book, which I have made use of, and which was subsequently reprinted unaltered! This singular maneuver was further clarified by a letter from Gérard Lebovici (Editions Champ Libre), dated 12 September 1980, on the subject of another French translation of *On Terrorism*[146]

---

[145] *Author's note*: Debord's *Preface* would appear along with a Dutch translation of the film script for [Debord's 1978 film] *In girum imus nocte et consumimur igni.*

[146] *Author's note*: There are two French translations of *On Terrorism*: one by Jean-François Martos, which I have made use of; and the other, which I haven't

that was sent to him in the hope of having it reprinted (a copy of this letter was sent to Sanguinetti).

As for the possibility of republication by Champ Libre, the comforting fact that the text has encountered a certain commercial success (as you have told me) has no importance here. Editions Champ Libre is entirely indifferent to all economic considerations, whether it is a question of gains or losses. And this is very fortunate, given the current centralization of book distribution, the servitude of the newspapers, the indigence of the bookstores, the boycott attempted from all sides, etc. (...)

Moreover, I have previously seen the complete manuscript of *Remedy for Everything.* The part that has since been extracted by the author and translated by you is incontestably the most interesting. I know that Gianfranco Sanguinetti merits esteem for the unique courage he has shown by affirming in Italy a truth that the powers-that-be [*des forces*] want to hide by every means possible. And I am happy that his words have caused many echoes in France and in many other countries, and will continue to do so in the future.

But in January 1976 I published the first non-Italian edition of *The Truthful Report,* which is an excellent and exemplary book. Naturally I cannot envision publishing a weaker and poorer book by the same author.

Sanguinetti deals with "the theory and practice of terrorism, developed for the first time" and clearly adds that his text permits his readers to "read it here, and only here." It seems to me that Gianfranco Sanguinetti's current firmness doesn't at all authorize his glorious tone on this aspect of the question. I myself published, in *February* 1979 a little book in which someone already said all of the truths that Sanguinetti published in April of that same year (this work was immediately sent to him and a translation of it appeared in Italy in May [1979]). What's more, I have photocopies of a correspondence exchanged while Moro was being held, still alive, between Sanguinetti and one of his foreign correspondents. This correspondent put him on guard by exposing the entire truth

---

seen, was published in Grenoble. I only received a copy of Lebovici's letter a few weeks ago.

of the affair, and advised him to reveal it as soon as possible. At the time, Sanguinetti responded by resolutely declaring his skepticism concerning this version of the facts, or he only pretended to be so for reasons that remain obscure to me. When one has lost several months before wanting to admit the obvious, there is something out of place in insisting on one's avant-gardist originality.

I find, therefore, that, from the point of view of Editions Cham Libre, the useful truths in *On Terrorism and the State* lack a bit of freshness.

We would be able to quite simply adopt this excellent position if this volume also included the *Truthful Report,* the two translations of *On Terrorism* and so many other books that Champ Libre could and wanted to publish; in sum, if, in this aspect, the conditions here [in Holland] weren't so different from those in France. The valuable arguments and the useful truths gathered together in *On Terrorism* apropos of the machinations to which the Italian State has had recourse, the decree of its decadence, and what it has done have been almost unknown here, until now.

We can only congratulate ourselves with what will henceforth be available to all those people who read Dutch and, besides, with what – thanks to this Postface – are not only revealed State secrets, but also the secret of their revelation.[147]

---

[147] *Author's note*: copies of this Postface have been sent to Gianfranco Sanguinetti, Guy Debord, Gérard Lebovici, Jaap Kloosterman ·and Jean-François Martos.

Letter from Sanguinetti to Khayati Concerning Debord

10 December 2012
Dear Mustapha,

I have reflected upon your hypothesis, according to which Guy, starting from a certain moment, "had the desire to minimize the role of his companions," and that, after his death, Alice has only been the executrix of this last will for effacement. It is not improbable that towards the end of his parabola, Guy had obeyed what one calls in Italian the "annihilation drive" [*pulsion d'anéantissement*], which brought him to annul the reality of the other and make it disappear as if it never existed.

In any case, before then and fortunately so, Guy wasn't at all like that, because I recall how often he had emphasized to me the important role of the first "artistic" period of the SI and, likewise, the considerable role played by this or that situationist, by saying – with the modesty of great men – that his merit had been his ability to grasp, solidify and give form to the impulses, thoughts, etc. that came to him from other people. Without minimizing his personal merits, this ability seems normal and unquestionable, because, otherwise, what's the use of a group? But this was the Guy that one loved, the one to whom one will always be grateful, whose works one admires and with whom one has been able to do the most beautiful things, and I in particular was able (and for a long time) to share a life together with him, of a richness that is rarely shared today, in the new conditions of the world. At a certain moment, things changed.

Your letter has caused me to take the effort to search through the correspondence and, moreover, through the documents that support your hypothesis. They exist. I haven't yet found a late letter (to an assistant on one of his films?) that asked him what became of the other situs, and Guy responded something like, "I made them disappear." But I recall reading that in the published correspondence.

In what concerns me personally, it is definite, in any case, that starting from the success of Operation Censor, in which he didn't believe very much, there came upon Guy a somewhat suspicious kind of caution towards me. And then he sought for a longtime a pretext to attack me, not frankly or directly, which was something he could do by writing to me directly, but obliquely, in an undeclared, asymmetrical war, and, a few years later, by spreading (left and right) insinuations and calumnious hypotheses about the Doge, about me, about my conduct in the Moro Affair, etc.[148] These practices, to the extent that

99

they were of a calumnious character, called for the only conduct to adopt against calumny: *to ignore it* and especially *not fall into the trap of defending oneself against it*. I did not respond to the manipulator or to those who were manipulated, neither then nor afterwards. This epistolary offensive reached its paroxysm in 1981, after the first two French editions and the editions published in German, Greek and Portuguese of *On Terrorism*, published in Grenoble, Paris, Hamburg, Athens and Lisbon – Dutch, English and Spanish editions were imminent.

Here I am obligated to make a long digression on something that concerns me particularly, which will allow me to better specify the strategy with which Guy proceeded, and what damage he proposed to do, because I have read the letters (many of them for the first time) that he wrote left and right during the summer of '81, when he was violently overtaken by a paranoid and maneuver-heavy crisis. One knows that paranoia is generally lucid, structured and systematic. With Guy, it filled out a letter addressed to Michel Prigent (dated 22 May 1981), in which, several lines apart, he wrote, "I am thus obligated, so that no one will have the occasion to say that perhaps I would manipulate this or that person, to break off all relations . . ." and then, "The *method of the truth* isn't a too-difficult application (...) I see a simple and clear example of it in Els van Daele's postface to *Terrorism*."[149] This postface, which was entirely or in part constructed by Guy, and which was imposed upon the Dutch publishers, *isn't the refutation* of manipulative practices, as he claims, but their *definitive and monumental confirmation*. Likewise for the "Foreword" imposed upon the English edition, which was signed by Lucy Forsyth.[150] Thus, I take the occasion of the present letter to you to get a few pebbles out of my boots.

It was at this precise time that Guy launched his offensive against me. The principal explication that I found for this murky operation is the success

---

[148] See my note on *The Doge*. [*Translator*: Sanguinetti's "The Doge: A Recollection" was written in December 2012 and shortly thereafter published in an English translation at http://www.notbored.org/The-Doge.pfd.]

[149] Cf. Guy Debord, *Correspondance*, vol. 6, p. 121.

[150] *Translator*: Sanguinetti's assertion that it was in fact Debord who wrote and imposed both the "Postface" to the Dutch edition and the "Foreword" to the English edition certainly explains the otherwise inexplicable "coincidence" that the two texts share, not only the same ideas, but the same way of phrasing them. Both texts also make use of the same excerpt from Debord's letter to Kloosterman dated 23 February 1981.

that my book on terrorism enjoyed abroad, where the people who published it or undertook to publish it did so with more impact than that produced by his *Preface to the Fourth Italian Edition of "The Society of the Spectacle"*. Guy thought that *On Terrorism* had more success than it actually had, the success of Censor still being fresh. His second line of attack aimed at eliminating me from the subversive movement by preventively discrediting all that I might still do and write, as well as what he had done with others.

These operations began with the sending of a dismissive letter, signed by [Gérard] Lebovici (12 September 1980), to Philippe Rouyau and Jean-François Labrugère, who were preparing to reprint their translation of my book into French. They continued with the recruitment of a bad soldier, the French translator Jean François Martos, in the spring of 1981, because it is from his translation that, generally speaking, the other editions of *On Terrorism* were made. By having Martos under control, Guy could also control the future editions, as was the case in Holland and England. He imposed a condition upon Martos, an *a priori*, as he called it, if they were to be friends and collaborators: his relations with me had to end. The hostilities subsequently continued with the distribution of four letters from three years before that, at the time of the Moro affair in 1978. They included the one from Guy (21 April 1978) in which he asked me to intervene in the affair and go underground (which would have been a very serious error in Italy at that time: anyone not found at his address was condemned in advance), and a letter from me (1 June 1978), in which I got muddled up in opposing hypotheses in order to gainsay the proposed strategy. I was wrong about the precise point, but I knew why, and this was neither serious, "suspect," "obscure" nor "guilty," as Guy would say three years later. Then there was another letter from me to Guy (15 August 1978) and the response from Guy (29 August 1978) in which he posed three questions to me.[151] My response (24 September 1978) to that last letter was not published. We will see why it wasn't.

Since life is (fortunately) richer and more complicated than Guy's paranoid simplifications, I will recount for you what I could not say to anyone. I had first-hand information from a German anarchist in the R.A.F. who was close to me at the time. I had previously seen her battle courageously at the barricades in Bologna in September 1977. Shortly afterwards, she immediately denounced the assassination of Andreas Baader, Gudrun Ensslin

---

[151] These letters were published in Champ Libre's *Correspondance* (vol. 2) in 1981.

and Jan-Karl Raspe in the prison at Stammheim. She was persecuted and quickly arrested by the Italian police at the behest of the Germans, who wanted to extradite her. She was also in contact with stray Italian terrorists of good faith; she had first-hand news; and thus, in my letter to Guy, I tried to envision things in the perspective of the information that I had come to learn in this fashion, without being able to discern how much of it was true or uncertain, nor was I (obviously cautious) able to cite my sources: in the Moro affair, the authorities had adroitly ensnared sincere terrorists who didn't know for whom they worked. Furthermore, you know that they had tried to implicate me,[152] though I'd never been a terrorist.

Thus, this was information that I could not communicate to anyone, and I certainly couldn't *write it down and send it* to Guy: the German anarchist – thanks to a famous marriage of convenience with the son of her Italian attorney a week before her arrest, which *ipso facto* made her an Italian – wasn't extradited to Germany, had narrowly avoided incarceration at Stammheim and had survived. I could not nor did I want to write to other people about this, not even in 1981. And especially not to a simple translator of my book who so impertinently demanded an accounting from me.

Martos, with whom I was never a friend, and whom I never considered to be a subversive, came to be – if not because of his vague desires, then because he'd translated *On Terrorism* – recruited for Guy's stable, and thus believed that he'd been admitted into the ranks of international subversion, when he was always nothing but an opportunist. He was commanded to write to me a letter that demanded an explanation for the contradiction there had been between the hypothesis advanced in my letter to Guy dated 1 June 1978 and the thesis supported in my book. Guy took pains to distribute my response to his three hypotheses,[153] written on 24 September 1978, *between two fire-bombings* of my house in the country. In that letter, I clearly rejected the first two hypotheses and partially admitting the third one. Although

---

[152] In the large-scale legal and police-related operation conducted by the anti-terrorist prosecutor Pier Luigi Vigna in November 1979, a few months after the publication of *On Terrorism*.

[153] "Thus I would like to know the reason that motivated those analyses, so strange, at the time: a) direct pressure from the authorities? b) indirect pressure from the same origin, but politely presented through the insinuations of the very suspect Doge? c) the pure pleasure of contradicting Cavalcanti, an activity to which you are devoted too often, to the detriment of better activities?" (Letter from Guy dated 29 August 1978).

truncated, the correspondence made public three years later was made to justify question marks.

Thus Martos sent me his eminently insolent letter (3 June 1981).[154] I did not respond. And for good reason. He had written his malicious and inquisitorial letter on instructions from Guy and he had written it to please Guy, and it demonstrated especially well that Martos had been recruited like a marionette. The goal of the letter was to set the bases for a subsequent campaign to defame and dishonor me in the poor circle of puppets with which Guy wanted to surround himself, a circle to which this letter was immediately sent. Here it is necessary to make clear that [Jaap] Kloosterman and [Michel] Prigent [who also received copies of Martos' letter] were not properly part of Guy's stable, but Guy was interested in them because each was preparing an edition of *On Terrorism*.

You know as well as I that, in Italy, I had done and risked, alone, what no one else at the time had risked to do in order to denounce *coram populo*[155] modern terrorism – which today one calls *false flag*[156] operations – and what I continue to risk every day. In the name of what would I need to satisfy the malicious and aggressive curiosity of parasitic spectators who, while I lived in a state of emergency in Italy, amused themselves by spreading ignoble suspicions against the only person they knew who had fought the enemy on the front lines? I had something other than their insolence, which bordered on calumny, to defend myself against. They could fuck themselves! Since I never responded to them, I pulled the carpet out from under their feet, and I am quite happy that I behaved in that fashion. They remained famished; I did not feed their hunger for explanations and gossip. *Never complain, never explain.*[157] They have nothing to be proud of.

In any case, twenty years later, on 11 September 2001, *it was my book that people recalled*, then already amply distributed on the Internet in several languages, and not theirs, who had inscribed their ignominious suspicions, their insinuations and their names in the infamous column that testified to their cowardice. They have never risked, neither then nor subsequently, producing anything efficacious against the new spectacular terrorism with which they filled their mouths. They preferred to accuse me of having denounced it *a little later* than they had desired. And if they complained of my

---

[154] Cf. J.-F. Martos, *Correspondance avec Guy Debord*, Paris, 1998, pp. 166-168.
[155] *Translator*: publicly.
[156] *Translator*: English in original.
[157] *Translator*: English in original.

supposed "lateness," and if they believed that it was truly urgent, *what the devil prevented them from preceding me?*

Martos was only a spectator and a puppet who has never been arrested or even interrogated by any police officer or judge; he has never experienced attacks, interrogations, trials, searches, or criminal charges – all the things to which I was subjected over the years.

At that time, in Italy as well as in Germany, one didn't have the leisure of masturbating with the spider webs of the French pro-situ opportunists. I had cops and saboteurs at my door, and my friends were arrested. Several died during those years. I had to defend myself against other dangers, quite real ones, imminent and threatening, not the dishonest provocations that came to me from the other side of the Alps. In 1979, Licio Gelli[158] activated the fiancé of one of his daughters, the son of a carabineer, to stage a provocation against me, after which I was indicted for the crime of contraband, and the affair ended with a violent encounter.[159] And it could have turned out worse.

I had neither the time nor the desire to confront a colossus the size of Martos. His correspondence with Guy is a catalogue of gossip. He has since then tried to sell it to the highest bidders and has sold none of it. His use value is identical to his exchange value. For all these reasons, it was wise for me to let him and the others talk. An old Tuscan proverb says: *"acqua che scorre e gente che parla non si parano."*[160] And so I let them run on.

Guy's strategy of attack was then deployed in numerous letters sent to several people, as one learns from his *Correspondance* (vol. 6). Therein he alludes to vague "serious reasons" and "reasons that have remained very obscure to me," to a "suspect attitude," to a "damning document" (?!) concerning my letter of 1 June 1978, with phrases that push his correspondents to believe in and let them imagine very serious things, but they are always shrouded in mystery ("I believe I have told you *the least possible* and, at the same time, the minimum necessary," he tells Martos on 24 July 1981) or even contain very hypocritical insinuations (in the same letter, he says, "I have been Gianfranco's friend. I certainly do not want . . . to discourage those who at the moment are his friends, by revealing to them all

---

[158] *Translator:* born in 1919, Gelli was a fascist and pro-Franco soldier in his youth. As an adult, he worked for the CIA, participated in the failed Borghese coup of 1970, and was a key member of NATO's Operation Gladio and the P2 Masonic Lodge.

[159] At the time, no one knew anything about Licio Gelli or Operation Gladio.

[160] "You can't stop water from running or people from talking."

that I know," *but without ever saying what that might be*). A little further down: "I want to warn you about certain dangers; I no longer know if Gianfranco knows about them or refuses to know about them," without ever indicating *what exactly* they might be, because – beyond a paranoid suspicion (or a pretext for suspicion) about the Doge, in either case *totally unfounded* and, furthermore, not stated in plain language – *there was absolutely nothing to say*. "I have told you to ask him what he thinks about the 'Doge' these days. This is a kind of password *to assure your protection*" (letter to Martos dated 24 August 1981). *Porca Madonna!*[161] Protection?! From what? Of whom? In Italy, it is the Mafia that offers protection! If there were dangers that I did not see, wouldn't I have the right to be immediately and directly informed? But these smoky dangers did not exist.

One could believe just about anything after these insinuations, which were as heavy as they were unfounded, and this was well and truly the reason for their existence, through which paranoia became contagious, as was the case with poor Carlos Ojeda, who indeed became crazy for a moment.[162]

To impose his postface (signed by Els van Daele) to the Dutch edition of *On Terrorism*, he wrote to Jaap Kloosterman, "One knows very well that Gianfranco *is guilty*" (letter dated 20 August 1981), because Jaap was a bit skeptical about what Guy had evoked so vaguely, and he was about to publish my book in Holland. And to Michel Prigent, who was about to publish an English edition, he wrote, "Sanguinetti has found nothing in the letter from Jeff [Martos] to respond to (...): which proves that Jeff's letter was strong enough to reduce to silence someone who is *so obviously guilty*" (letter dated 29 August 1981). And then, that same day, in the grips of the same acute crisis, he wrote to Carlos Ojeda: "One knows very well that Gianfranco has been *guilty* for a long time and in the eyes of many people, due to what he hasn't said and what he has said." Because he kept himself from saying precisely *what* I was guilty of, one might have well and truly believed oneself to be in Kafka's novel *In the Penal Colony* (*In der Strafkolonie*), in which "the crime is never in doubt." In any case, one knows quite well that calumny surpasses any demonstration: it is content to be repeated and passed from mouth to ear.

Once again, in the same paranoid attack of the summer of 1981: "Gianfranco hasn't responded to you (...) This is a terrible verification: even more than what I could have thought" (letter to Martos dated 29 August

---

[161] *Translator*: Holy fuck! (literally "Pig Madonna").
[162] Cf. Guy's letter to Carlos Ojeda, dated 29 August 1981.

1981). Verification *of what?* Not *of the truth*, certainly! What *was verified here was only his obsession.*

Here Guy comported himself in a dishonorable manner, one worthy of some politician: he knew me better than anyone since 1969, and he knew well that I always comported myself in an inflexible, courageous, adroit and irreproachable manner with all the authorities, police officers, members of the army, judges, ministers, provocateurs, and jailers, and this in all the different countries in which I had to deal with them. It was this very comportment that always saved me. And it was thanks to my comportment in such dramatic situations that I was able to save many people from even worse consequences.[163] Guy didn't even fear to write as one might chronicle diverse facts in a newspaper for the plebs: " . . . since Gianfranco, in whom this genre of detestable acumen hardly surprises me, has managed to not respond [to Carlos Ojeda] on all the burning questions (...) what consequences shouldn't one fear?" (letter to Martos, 29 August 1981).

Here I note in passing that the Dutch and English editions of *On Terrorism*, both published in 1982, are the most striking examples of schizophrenia in the history of publishing since *Anti-Machiavel* by Frederic II and Voltaire.[164] Both of these editions publish my text and, at the same time, launch an attack against my person (under the signatures of Els van Daele and Lucy Forsyth). This gives the impression that the book was only published so that their suspicions about and censures of its author could be spread.

The height of the hypocrisy and false consciousness was only reached two months later, when Guy, feigning scandalized innocence, wrote the following to Lebovici on 18 October 1981: "Do you know the most recent book by [Jean-Pierre] Voyer? In it he incites [the reader to] the murder of Sanguinetti by insinuating that he has deliberately worked for the Italian police." Voyer, who had always been crazy, had published a manifesto in which he said, among other things, "The question that is justly posed with respect to Sanguinetti is: how is it that he is still alive and free? After all, he is

---

[163] Where this is concerned, the comportment of the French Minister of the Interior (Raymond Marcellin) – always well informed by his cops – seems to me more astute and realistic when he declared the day before my expulsion from France (and thus tried to justify it), "It is a very poor tactic to pretend to propitiate an irreducible adversary by increasing one's concessions to him (...) To come to terms with him is to add shame to an assured defeat." (*Le Monde*, 27 July 1971).
[164] *Translator*: a chapter-by-chapter rebuttal of *The Prince* published in 1740.

perhaps an agent of the secret services. But one has seen that such an affiliation is no longer a guarantee of security in Italy today." But despite his madness, Voyer honestly added: "It is necessary that the author of the *'Protest'* applies to himself the method that he applies to Sanguinetti. One must grant Sanguinetti the merit of having denounced the spectacular usage of Stalinist terrorism in Italy that is made by the Italian State" (Jean Pierre Voyer, *Response to the Author of 'Protest to the Libertarians of the Present and the Future Concerning the Capitulations of 1980',*[165] Paris, 7 October 1981).

Faced with Guy's ingenuous, outraged innocence concerning Voyer's poster, one must wonder: *what had he himself insinuated that was different just the day before,* but with more authority and emphasis? Wasn't he the first one to have appealed for a kind of virtual semi-public lynching of me? After seeing Voyer's manifesto, had he finally perceived the effects of and reactions to his calumnious allusions, only two months after making and spreading them around?

Whatever the case may be, all this is pitiful, and borders on ignomy, or at least an explanation in terms of pathology. In an extremely virtuous declaration, made hardly a year prior to that, in a letter to Diego Camacho (written by Guy but signed by Lebovici) that criticizes Camacho for his insinuations about the death of Durruti, Guy proclaims: "Champ Libre never publishes authors who insinuate, whatever the political necessities they support"![166] And he wrote to Lebovici on 3 July 1980: "Obviously you cannot publish Camacho, who *insinuates* (...). He even insinuates with a poisonous caution (...) He boasts that he 'insinuates'. . ."

---

[165] The pastiche called *Protestation devant les libertaires du present et du futur sur les capitulations de 1980,* which caused so much ink and suspicions to flow among Guy and the others in his circle, was in reality written anonymously by someone named Jean-Claude Lutanie, who died in 2006. It was reprinted in 2011. [*Translator*: in the words of the editors of the 2011 edition: "The *Protest* above all testifies to a disappointment with respect to situationist thought and its 'youthful, unkept promises.' Lutanie essentially takes aim at Debord, accusing him, if not of jealousy for a radicalism of which he was quite incapable, although he claimed it, then at least bad faith in his affirmation that the group [Action directe] was manipulated by the State." Cf. Debord's letter to Kloosterman dated 13 July 1981: "I have asked Gérard [Lebovici] to send you a copy of a short, very shady pamphlet (...) which, it seems to me, must be read with great attention."]

[166] Cf. Champ Libre, *Correspondance*, vol. 2, p. 63, letter dated 4 August 1980.

Thus, using his own insinuations and aiming at his own particular goals, Guy positively (and vainly)[167] tried to make me disappear by defaming me. But *he failed* because he never succeeded in convincing anyone other than his sycophants, and he also didn't succeed in breaking up [*franchir*] the bigger circle of the revolutionaries, in Italy and elsewhere, who knew me, that is to say, those who [in his eyes] merited the title "fanatical, inept Sanguinettists."[168] The actions against me taken by the judges, police officers, provocateurs, Stalinists and fascists at least had a justification in the damage that I had done them and would continue to do them.

It is certain that, for the man who wanted to be a strategist, who admired Clausewitz and invented *Kriegspiel*,[169] the campaign against me in the summer of 1981 essentially ended in a single result: the precious acquisition of an authorized "historian" of the SI.[170] I believe that Guy must have realized this, because this campaign was very quickly ended and abandoned, and no one spoke of it any more, as if nothing had happened, which is also quite strange. His serious and quickly forgotten suspicions sounded false and became suspect.

One can circle around this problem as much as one wants: it remains that, from the strategic point of view, this campaign was a failure and even a reversal: neither the English, the Dutch, the Germans, the Greeks, the Spanish, the Portuguese, the Italians, nor (later on) the Americans joined the small French pro-situ sycophantic herd that was thus reduced to Martos alone, who was unanimously scorned. But the greatest defeat for Guy was that, for once in his life, he was cut down to the role of a manipulating and maneuvering politician. I am happy that I didn't have any role in this low-down game, that I didn't feed it, and that I remained impassive in the face of the calumny.

It is true that, in 1981, I didn't have all the documents that are available today. In them, one can even learn that one of the Frenchmen who came to my place in the country reported to Guy, either directly or by way of [indirect] gossip. But even if I had known all of the hidden agendas of these dealings, my quite anti-politician character would have prevented me from standing up in such an arena. In any case, I knew enough about them at the time to choose to

---

[167] *Translator*: both due to his vanity and unsuccessfully.
[168] Cf. Debord's letter to Martos dated 25 February 1982.
[169] *Translator*: a cabinet game marketed by a company formed by Debord and Lebovici in 1976.
[170] *Translator*: Martos' *Histoire de l'Internationale Situationniste* was published Editions Gérard Lebovici in 1989.

let the enemy fleet, which was so badly piloted, run aground against the reefs that broke its reputation, without lowering me to their depths.

In that climate of poison, disinformation and hostile and dishonest intoxication, Guy still dared to complain that I'd never cited his *Preface to the Fourth Italian Edition of "The Society of the Spectacle"*. He wrote the following to Lebovici on 25 June 1980: "Thank you . . . for the French edition of Gianfranco . . . The preface for the French edition is, in any case, better than the rest of it. It is true that the author doesn't cite me, but who ever cites me?" Perhaps that was my real capital crime! But, moreover, the question itself was badly posed: *it was him* who should have cited and even supported me – knowing [as he did] that I was fighting behind enemy lines, in extreme conditions – because, when he wrote his *Preface* (January 1979), he had already read the manuscript of *On Terrorism* (October 1978),[171] which had been written well before his text. It is also true that, in May 1981, he felt the necessity of defending himself against a suspicion *that no one had raised* when he noted the following in his postface (signed by Els van Daele) to the Dutch edition of *On Terrorism*: "Since there are a great number of concordances between the two writings, from the choices of historical examples to certain stylistic details (...), the pages of the *Preface* (...) appear to the reader as a summary of *On Terrorism*." This clarification has the air of an *excusatio non petita*.[172]

Why did Guy place himself on a slope that was so slippery for him, when there was no necessity to do so? That's the question. In his politician-paranoid downward slide [*dérive*], there was – as is often the case in paranoia – a method; in any case, there was a paralogic that satisfied itself with the pseudo-reality that it itself had created in order to combat *an effigy* of a pseudo-enemy. *False flag*[173] terrorism, or religion, functions in the same manner, as one knows. On the one hand, Guy recognized the arguments I made in *On Terrorism*; on the other hand, he wanted to create an emptiness around me, and he tried to stop any influence that I might have in the subversive milieu of the time, not because he thought that my influence was harmful, but, on the contrary, because he recognized that my book "is very

---

[171] *Translator*: cf. Debord's letter to Paolo Salvadori dated 12 November 1978: "I have read Gianfranco's manuscript."

[172] *Translator*: Latin for "an unprovoked excuse." Usually this phrase precedes the words *accusatio manifesta*, with the entire sentence meaning "an excuse that has not been sought [is] an obvious accusation" (against oneself).

[173] *Translator*: English in original.

true, and [it would be] very good to have it be known, as soon as possible, for its exact analysis of the Red Brigade" (cf. his letter to the Greek Mikis Anastasiadis dated 25 June 1981). And, as he wrote to Kloosterman on 23 February 1981, "I think that it is a very good thing to publish *On Terrorism*, which is exactly true on the central question that it concerns, and full of very valuable arguments about it." A year before that, in a letter to Anastasiadis dated 5 August 1980, he wrote, "You no doubt know that [to answer the demand for it] Gianfranco's book has already had a second French edition. Thus pseudo-terrorism begins to find its antidote . . ." Guy's contradiction was thus between the affirmed utility that he recognized in my book and his desire to diminish its importance or to make me disappear.

Did Guy believe that it was my influence or that of the other situationists that was diminishing? Or quite simply, did he not want me, as one says in Italian, scratching around and going through his henhouse? And this was a man who knew very well that Retz had already established that, in matters of calumny, everything that doesn't do harm works in favor of, not against, the one who is attacked.

This attitude, reinforced by the "annihilation drive" of which I spoke at the beginning of this letter, naturally was extended to all those who had contributed the most to the subversion of which the SI was the bearer. He wanted to remain alone. Thus it is not an exaggeration to say that, from that moment on, he began to systematically minimize the role played by all the [other] members of the group. As a result, there was only room around Guy for mediocrities and opportunists, whom he launched, in an adventurist manner, against those who had been excellent. This has had obvious consequences, even after his death, in the great work done by Alice (I say "great" in the sense of volume). One of the consequences of this degradation are all the pathetic books written to the glory of Guy: the sycophantic biographies and the supervised pseudo-histories of the SI by a multitude of revisionist "historians" and by impecunious "philosophers," professors, journalists, etc., who have allowed themselves to be shamelessly herded, flogged and censured by Alice. There has also been a proliferation of publications by essayists, archivists, and laborious and impoverished opportunists seeking the misplaced vanity of the backseat driver. None of this has happened by chance: it has been desired and promoted by Alice, but, before that, by Guy himself.

Among the apologists one can find real pearls: for example, the book by Apostolidès, which, in the fury of making me disappear, reaches philological summits never attained even by the KGB. After having claimed that the French version of the *Truthful Report on the Last Chances to Save Capitalism in Italy*[174] was more "elegant" than the Italian original (!), and in order to complete his demonstration that Censor was not Sanguinetti, but Debord, he removes all doubt with the following wise lesson: "One remarks the affinities between the two names, Censor and Debord: they each possess two syllables; the vowels are identical, as are the number of letters."[175] The "affinity" for which I chose the pseudonym *Censor* is, on the contrary, with *Bancor*, the supranational currency invented by Keynes; it was also the penname of Guido Carli, who was the head of the Bank of Italy at the time. That is quite far away from the furious demonstrative keenness of Apostolidès, the unfortunate orphan of Pope Pius XII,[176] Mao and Lenin who only demonstrated that his spastic research was part of a spectacular cult of personality.

To tell the truth, I must confess to you that I wasted very little of my time documenting the above (except for the book by Jappe,[177] which, at the time it was published in Italy, I warmly recommended to Alice that she publish in France: but it too, if I recall correctly, was published with some censorship or arrangements). The fact is that reading such works immediately bored me; they have no historical value because they are all stale products due to servility or Alice's directives. The real struggle *is completely absent from them*, to the profit of a mythological and laughable combat that is designed, after the making of historical revisions, to have this or that individual appear or, rather, disappear (preferably to make all of them disappear) so as "to certify that he (Debord) remains the Unique One."[178] I

---

[174] *Translator*: it was Debord who translated the book from the Italian into French.

[175] Jean-Marie Apostolidès, *Les Tombeaux de Guy Debord*, Paris, 1999, cf. 99-104.

[176] *Translator*: the infamous pope who reigned during World War II.

[177] *Translator*: Anselm Jappe, *Debord* (Pescara: Tracce, 1992).

[178] Apostolidès, *Les Tombeaux de Guy Debord*, p. 103. Another official falsifier, even more mean-spirited (if that is possible), is Vincent Kaufmann, who concludes, "The SI is to be considered as the work (in all the meanings of the word), or as one of the works, of Debord alone" (*Guy Debord: La Révolution au service de la Poésie*, Paris, 2001, p. 278). This judgment is repeated again, on p. 285, to better convince the reader. On p. 277, Kaufmann writes, "The Italian Gianfranco Sanguinetti (...) was nothing in it . . . he was truly not in the loop during the final debates that shook the SI." One hundred pages later, he continues, "Behind Censor, there was indeed Sanguinetti, but behind

have been contacted several times, fortunately bashfully, by these almanac salesmen, to whom I have never given anything to nibble upon. There would have been much to discuss about these productions, if I didn't have anything better to do, or if I instituted a pricing structure for the most boorish of these authors: Bourseiller,[179] Martos, Kaufmann, Apostolidès or another seller of Bordeaux wine.[180] They will be quickly forgotten. Thus, let us move on.

The most surprising thing is that the practices cooked up by Guy for his sycophants, in addition to his arbitrary [*occasionnelle*] versions of things, have been blindly taken up, without any verification or documentary research, by these alleged historians. This crude and simple-minded ideological vision allows these brave militants of history to share and propagate a mythological and weak-willed version of the facts. It is passably comic to ascertain their quasi-unanimity on the idea that Guy's departure from Italy (among other things) was caused by a shadowy persecution or expulsion of which he had [supposedly] been the victim in 1977. In fact, his departure, which took place before the Censor project, was the consequence of a series of prosaic facts: disappointments with Florentine girls; irritation with the fact that he had cut off the gas-heating in Florence in mid-winter; someone had stolen the wine from the cellar of the priest of Pieve de San Cresci, where he'd been staying; banal monetary problems that I had had, etc. Thus, there are a crowd of errors of this sort on the tombstones[181] and monuments erected to his glory, all of them working upon the fabrication of a legend and myth.

The first wave of makeshift "historians" has been merrily burned and sacrificed upon the altar of sycophantic praise, which Guy, quoting [Jonathan] Swift, liked to recall is the daughter of existing power. If he got wind of these tombstones, I believe that he would rather have concluded with the words of Schopenhauer: "That soon the worms will nibble on my body, this is a thought that I can tolerate; but the idea that the professors will do it with my philosophy, that horrifies me."

Sanguinetti was Debord, or at least his style, in all the senses of the term (...) Is it so surprising that he hastened to translate the book by Censor into French?"
[179] Bourseiller succeeded in the brazen enterprise (among others) of writing an essay titled "L'IS face au Terrorisme," without citing me, if only to state that the *Truthful Report* was written by Guy (cf. *Archives et Documents Situationnistes*, 2, Denoël, Paris, 2002).
[180] *Translator*: an allusion to Philippe Sollers.
[181] *Translator*: rendered into English, the title of Apostolidès' book is *The Tombstones of Guy Debord*.

I hope, Mustapha, that this letter will serve to throw a little light on the self-interested confusion that surrounds the distribution of my book on terrorism.

Best wishes,
Gianfranco

Index

www.ingramcontent.com/pod-product-compliance
Lightning Source LLC
Chambersburg PA
CBHW031520270326
41930CB00006B/454